Chris,
Thanks for all your support.
You are great friend and co-worker
Boyd
"MR. SUMMERS"

Effective Methods for Software and Systems Integration

Effective Methods for Software and Systems Integration

Boyd L. Summers

CRC Press
Taylor & Francis Group
Boca Raton London New York

CRC Press is an imprint of the
Taylor & Francis Group, an **informa** business
AN AUERBACH BOOK

CRC Press
Taylor & Francis Group
6000 Broken Sound Parkway NW, Suite 300
Boca Raton, FL 33487-2742

© 2013 by Taylor & Francis Group, LLC
CRC Press is an imprint of Taylor & Francis Group, an Informa business

No claim to original U.S. Government works

Printed in the United States of America on acid-free paper
Version Date: 20120326

International Standard Book Number: 978-1-4398-7662-6 (Hardback)

Library of Congress Cataloging-in-Publication Data

Summers, Boyd L.
  Effective methods for software and systems integration / Boyd L. Summers.
    p. cm.
  Includes index.
  ISBN 978-1-4398-7662-6 (hardback)
  1. Computer software--Development. 2. Software architecture. 3. Systems integration. 4. Software engineering. I. Title.

  QA76.76.D47S86 2012
  005.1--dc23   2012003054

Visit the Taylor & Francis Web site at
http://www.taylorandfrancis.com

and the CRC Press Web site at
http://www.crcpress.com

Contents

List of Figures

List of Tables

Preface

I have been motivated for years to write this book, *Effective Methods for Software and Systems Integration*, due to integration challenges in military and aerospace programs and software industries. My previous software engineering book, *Software Engineering Reviews and Audits*, provided the framework and detailed requirements for verifying and performing audits during software design/development efforts. Performing reviews and audits that are successful ensures compliance in specified requirements, software design, testing, released configuration baselines, formal audits, and customer satisfaction.

The military and aerospace programs and projects that design, build, and test software work products effectively provide the framework to receive subcontractor and customer contracts. Opportunities to work in the technology field of software design/development provided me the perspective and understanding of day-to-day software engineering activities. To have effective software and systems integration methods in place provides an understanding of the importance of planning, coordination, software design, configuration management, integration, testing, subcontractors, and quality.

It is critical that integration schedules are addressed and coordinated daily with affected software teams and organizations before delivery inside software and systems integration environments. The software design/development life cycles must be completed and configured before baselines are released for test, integration, and functional checkouts.

Effective Methods for Software and Systems Integration delivers quality work products on schedule to customers.

SUMMARY

It is critical to understand and implement the disciplines during the software design/development life cycle prior to deliveries of software baselines inside software and systems integration environments. Chapters in this book define methods for dealing with project planning, systems design, software requirements, software design, software implementation, software integration, software and systems integration, subcontractors, delivery, and product evaluations to produce quality work products. *Effective Methods for Software and System Integration* will benefit current and future military and aerospace programs and projects.

Acknowledgments

I want to thank my lovely wife, Jana, for her support while writing this book.

My current and past software managers, software and system operations managers, and the software teams are an inspiration. They have given me the opportunity to be a contributor and a team player inside software and systems integration environments.

About the Author

Boyd L. Summers is currently working as a software engineer for the Boeing Company in Seattle, Washington. With 30 years of experience in software engineering and as a leader of multiple software development teams, Boyd continues to solve complex technical challenges to ensure that system and software engineering problems are addressed, resolved, and compliant.

Boyd is also the author of the software technology book, *Software Engineering Reviews and Audits*.

For questions about current and future software technology solutions, e-mail bl.summers.llc@gmail.com.

1

Introduction

The primary purpose for the implementation of *Effective Methods for Software and Systems Integration* does increase communication, knowledge, visibility into the software life cycle, and the importance of integration operations. Readers will find this book informative and interesting and will convey the methods for software and systems integration to be more effective in developmental military and aerospace programs and project development. The software industry/companies could benefit as well by adopting these effective methods. Enjoy the book.

1.1 SOFTWARE AND SYSTEMS INTEGRATION METHODS

To develop, operate, and maintain software and systems integration capabilities inside work product facilities, there must be a major discipline in supporting the entire software life cycle (i.e., planning, systems, requirements, design, builds, installations, integration, subcontractors, quality, and delivery) that does need to be completely understood. The critical understanding and the start of the right disciplines of these methods will empower and achieve effective, flexible, and quality results in an integration environment. The right disciplines are identified in Figure 1.1.

Effective: In the software industry/companies, military and aerospace program and projects do become effective by the implementation of achievable schedules, sound processes, and working solutions for software and systems integration.

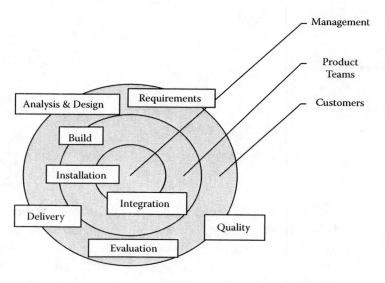

FIGURE 1.1
Start with the right disciplines.

Methods: Provide effective methods to ensure processes and tools improve productivity and prepare for the challenges that have an impact on integration environments.

Software: Software design, code and unit test, plans, and test procedures integrated with applied systems, tell us that the software developed is done right. "Peer reviews" are key.

Systems: Accomplish allocation of software design and engineering practices for systems to be defined and documented as ready for the combination of software and systems integration.

Integration: This is the compass to combine software, systems, firmware, and hardware to work together as one.

Rock Solid

This slogan is a reminder of the hard times and trials we face and the experience while software and systems talk to each other and improve software design and development efforts.

TABLE 1.1

Planning and Engineering Task

Software Engineering Tasks	Communication	Planning	Risk Management	Deployment
Systems design	x			
Requirements		x		
Design		x		
Configuration control			x	
Integration			x	
Delivery				x
Subcontractor				x
Quality product evaluations				x

1.2 PROGRAM AND PROJECT PLANNING

The purpose of program and project planning is to provide the necessary process steps to scope out planning for systems and software design/development within integration efforts. This type of planning will ensure and establish effective plans and results for performing the disciplines for software design/development, processes, and procedures for the implementation supporting software and systems integration activities. The planning and engineering task presented in Table 1.1 explains the disciplines and methods pertaining to communication, planning, risk management, and deployment.

1.3 SYSTEMS DESIGN

The method for systems design is to analyze customer requirements and develop a software design/development migration plan for defining the architecture, components, modules, interfaces, and necessary data for a designed system to satisfy specified requirements. The systems design method is increasingly important as it provides the disciplines required and implemented during software design/development life cycles.

1.4 SOFTWARE REQUIREMENTS

Defined and documented software requirements provide a systematic approach to development from multiple resources. The results of functional software interfaces, performance, verification, and production with required plans, documentation, and procedures become the basis for software design or development. This effective method is applied for initial development of software requirements and changes to requirement baselines.

1.5 SOFTWARE DESIGN/DEVELOPMENT

The software design/development definition is that of a systematic approach for the creation of software design and its development to reflect design and software definitions applicable to the work product. The resulting method defines details about the work product architecture, behavior, components, and interfaces. The software requirements are established between the elements of the design/development. The documented program and project plan provide traceability according to software-defined processes and procedures.

1.6 SOFTWARE IMPLEMENTATION

The importance of software implementation is a requirement for informal and formal integration testing in a development, integration facilities, or the software systems integration facility (S/SIF). The software implementation method for testing provides assurance that engineering builds function as expected to enable smooth execution for verification and test activities. An incremental software and test approach adds the functions incrementally in a series of engineering builds. The software design/development matures through a series of engineering builds. As software is tested and demonstrated, build plans are modified for subsequent builds based on lessons learned from previous engineering builds, troubleshooting, and checkout.

1.7 SOFTWARE INTEGRATION

All software delivered or implemented by software integration or testing is processed through a configuration and controlled software library system that maintains the official status accounting for each delivery. The integration tasks require that software design/development and test processes be in place to ensure integration is ready for team troubleshooting and checkout.

1.8 SOFTWARE AND SYSTEMS INTEGRATION

The software and systems integration method provides a consistent approach to effective integration activities. The software units, components, and subsystems are assembled in accordance with defined and documented integration procedures to ensure that the software and systems elements are assembled properly. The integration levels and the development plan (DP) for software determine if constructed elements are ready and subject to verification or validation activities.

1.9 SOFTWARE SUBCONTRACTOR

The software subcontractor roles and responsibilities describe how a program and projects can benefit and rely on outside companies' resources to provide required software and hardware products to be under contract and effective. The subcontractor presentation to the customer must be understood from the start of the presentation to the end. Questions can be asked by the customer to ensure that answers meet the needs for reliability and support.

1.10 SOFTWARE AND SYSTEMS INTEGRATION DELIVERY

When it is time for software and systems integration delivery, the delivery requires integration testing to be performed to provide assurance that both software and systems are integrated and working together. The integration practices ensure that units tested are complete and documented prior to the official delivery for the customer.

1.11 PRODUCT EVALUATION

An effective product evaluation method provides the necessary process steps to conduct and perform continuous evaluations of software work products during the design/development life cycle and integration activities. Numerous product evaluation tools and checklists are developed with associated scheduled processes to perform required audits and evaluations.

1.12 CONCLUSION

Defined software disciplines include an approach or method during the software life cycle for a program and projects to provide a plan from start-up to final delivery to the customer. Many methods were discussed but the number one method is "quality first"; the other methods come in second and so on as illustrated in Figure 1.2.

Many program and project meetings are called by senior managers; in attendance are software and hardware engineers. In those types of meetings, the hardware teams will sit on one side of the table, and the software team will sit on the opposite side. That situation is unique, but that is how it is at times. The senior managers and program and project managers attend meetings and discuss with the two teams that the software or system is not working correctly when it is time for delivery to integration facilities. There is finger-pointing, and both teams may blame the opposite team for the problem.

The senior manager then points to the program and project managers and says, "Fix this problem." That is why effective methods for software and systems integration need to have hardware and software designers working together to solve issues that could have an impact on integration, quality, and delivery schedules to the customer.

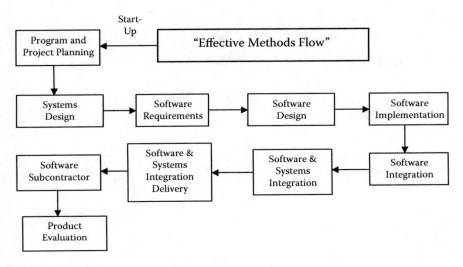

FIGURE 1.2
Effective methods flow.

FURTHER READING

Appleton, B., 2000. Patterns and software essential concepts and terminology. http://www.cmcrossroads.com/bradapp/docs/patterns-intro.html.

Beck, K., 2002. *Test-Driven Development: By Example*, 2nd ed. Addison-Wesley, Boston, MA.

Curritt, P.A., M. Dyer, and H.D. Mills, 1986. Certifying the reliability of software. *IEEE Transactions on Software Engineering*, SE-12(1), 3–11.

Gibb, T., 1988. *Principals of Software Project Management*. Addison-Wesley, Boston, MA.

Martin, R., 2000. *Engineer's Design Principals and Design Patterns*. CRC Press, Boca Raton, FL.

Phadke, M.S., 1997. Planning efficient software tests. *CrossTalk*, 10(10), 11–15.

2

Program and Project Planning

2.1 INTRODUCTION

Program and project planning is important as it describes the necessary planning for software and system efforts during software design/development life cycles. The definitions of systems design, software requirements and design, configuration control, systems and software integration, subcontractor involvement, deliveries, and product quality evaluations are critical to effective planning efforts. The initiation of planning starts at the proposal phase with the customer. The result of defined software design/development plans, processes, procedures, subcontractor support, and effective software tools provides estimations for cost and schedules to be available for teams that are impacted from the start of the proposal phase to delivery of the work products to the customer.

2.2 PROGRAM

Before a program can require a plan, program objectives are defined and technical and management disciplines are identified. This information defines a reasonable estimate or cause for:

- Cost evaluations
- Risk management assessments
- Defined and documented tasks

- Manageable schedules
- Progress reports

The people who work in a software design/development life cycle are expected to meet and achieve program objectives and understand the high expectations required of them. These activities begin at the systems design level of engineering and flows down to other software disciplines.

Program objectives identify goals for the program with consideration of how these goals are to be accomplished. Effective programs that perform to defined objectives and within the scope are successful due to implementing:

- Required data
- Tasks or functions
- How the work product performs
- Quantitative mechanisms

When program objectives and the scope are considered, program managers can select the best approach that would eliminate "roadblocks" imposed by scheduled delivery deadlines, budget concerns, and people issues.

2.2.1 Framework Established

Software processes provide the framework and effective planning when it is time for deliveries to software and systems integration facilities and the customer. Activities related to the framework are required for all programs. For effective planning, there are multiple tasks, scheduled milestones, and quality aspects necessary to ensure the framework is established not only for managers but also for the people who work. The quality assurance team, which is independent at times, and configuration management personnel monitor the framework processes.

2.3 PROJECT

The main reason that software projects are planned and controlled is to eliminate any confusion that could occur. The teams that are expected to provide work products struggle if projects are not planned, and control is not even an option. Studies showed that when schedule, cost, and quality objectives are not a top priority, the project is not successful. Although project success rates are improving, the failure rate can be high when an objective such as quality attributes is not implemented. To avoid failure, the project manager and a team of systems and software engineers who build work products must develop an approach for project planning, oversee activities, and ensure configuration control is in place.

Software projects get in trouble when uncertainty and confusion come into play. There are times when systems and software designers do not communicate, so defined requirements are not discussed in relation to the developed work product. To eliminate this lack of communication, guidelines must be established, such as:

- Structure daily meetings
- Share ideas
- Inform project managers of problems occurring
- Listen and try to resolve complaints

Projects can have a "daily standup" meeting. One a day can get to the critical points or problems and resolve them that day. If there are concerns, discuss issues with a project manager, then you do not waste other team members' time. In the past and currently, these meetings have an impact on hours of work that could be accomplished. My philosophy is that project managers need to have the confidence that their people can take care of the daily routines, so the project managers do not need to attend meetings for hours and hours. Time is lost; then, people are ready to go home for the day knowing they have time sitting around listening to people who have no impacts on what they are trying to accomplish that day. Stop this right now. Let us go to work.

2.4 PLANNING

Communication planning principals define goals and objectives during the course of program and project planning. The planning aspects require a set of managers to understand not only their position but also the technical practices that support systems and software engineering and to define the course that lies ahead. There are many planning ideas and decisions by managers that are not accepted by team members due to the complexity of change. What should you do? Under planning, the program and project should consider eliminating chaos. The pressure on teams can be enormous, and useful guidance can be provided, such as:

- Providing a scope for the team to know what is ahead
- Involving systems and software teams to help with delivery schedules
- Planning to adjust and accommodate change
- Identifying risks that could have an impact on program and project planning
- Defining and understanding quality
- Tracking the progress daily and adjusting if needed

2.5 SENIOR MANAGEMENT

At the senior management level, program and project managers are required to provide effective planning and focus so teams can be effective during software design/development activities. Failure in planning is not an option and does jeopardize the success in achieving sound practices in program and project execution. Communication early in the process is the key to eliminate risks and the ability to embark on operational deployments. The required job of a senior manager is to provide the common framework for program and project planning to address engineering tasks.

Many software managers begin their careers as software designers or developers. These types of managers serve:

- The company, military, and aerospace program and projects
- Their employees
- Themselves

When a software manager's team or organization delivers software to a customer in a timely fashion, this is called *execution*. There are questions that involve execution, such as:

- Do you have customer requirements?
- Do you have an approved budget?
- Do you have an approved plan and schedule?
- Are your program and project capable of dealing with change?

- Do you keep everyone focused?
- Do customers encounter quality issues with delivered work products?
- Do you measure work status on a regular basis?
- Do you find ways to improve?

Communication is important. A good software manager must learn to communicate in different ways, for example, providing formal presentations for upper-level management. Face-to-face communication to explain agreements with other program and project managers provides a road map and the plans for meeting goals.

E-mails work at times, but having a discussion will open up your team members to explain good and bad news. Also, communication is a positive way for team members to understand your expectations.

Program and project schedules that are not understood from the start will have an impact on resistance. To implement and use unreasonable schedules will imply that organizations and team members are not working hard. Customers are best served by creating work products that can be used over a long period of time.

Software managers must be aggressive and demand the best from designers and developers, but do not abuse them. Manage your teams wisely.

2.6 PROGRAM AND PROJECT PLANNING

The program and project planning method is well defined in the project planning process area in CMMI® for Development (version 1.3). The process area states the following:

> The term "project plan" is used throughout this process area to refer to the overall plan for controlling the project. The project plan can be a standalone document or be distributed across multiple documents. In either case, a coherent picture of who does what should be included. Likewise, monitoring and control can be centralized or distributed, as long as at the project level a coherent picture of project status can be maintained.

The scale of numerous software design/development efforts is huge and can lead to confusion and coordination with affected teams. Internal organizations in programs and projects develop schedules and define processes and tasks. At the senior management level, managers assign responsibility, authority, and accountability to program and project managers or team leaders to define the software design/development (i.e., systems and software design, configuration management, quality engineering, etc.) to provide required support.

Planning activities include:

- Software lessons learned from previous programs and projects
- Cost and schedule estimates and staffing plans
- Software and system requirement definitions
- Defined safety and security requirements
- Selection of appropriate software subcontractors
- Engineering documentation and historical data impacts
- Program and project objectives
- Contract understanding of required or necessary requirements

2.7 PLANNED SCHEDULES

The planned schedule defines tasks and processes to be conducted for implementation of those tasks and processes. The schedules that are planned affect team capabilities for risk assessment, configuration control, and quality. There are three critical factors in many software design/development programs and projects (Figure 2.1). The scope, schedule, and budget combined affect the quality of work products.

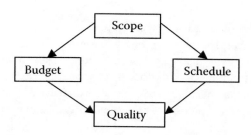

FIGURE 2.1
Planned schedules.

2.8 DEVELOPMENT PLAN

The critical items pertaining to a documented development plan consist of planned schedules and provide engineering information and direction for the production of software. It is important to know that the planning process is consistent with system-level planning. All major software design/development activities require consistency in accordance with the steps outlined in the use of development planning, including the following:

- Definition of entry and exit criteria for the software design/development
- Review and assessment of the work product and task requirements
- Definition or updates of the process for each software activity
- Development or update of the estimating process
- Development of initial cost and schedule estimation and risks
- Preparation of detailed implementation plans

2.9 TEAMWORK

An important element in all software programs and projects is *teamwork*, the coordination and communication within teams applied to meet work expectations. The effective methods for systems and software planning coordination provide value for a program and projects to far exceed high expectations. The software design/development energy and consistency appeal to achieve high-performance goals and aspirations. By having trust among teams, a cohesiveness is maintained in the work environment, and planning schedules becomes much easier to coordinate and implement within the team.

A plan developed is correct or successful when the team delivers a high-quality work product on time to meet the schedule and works within the budget. Remember that senior managers must encourage the program and project managers to work together with their teams to become effective, respond to customer expectations, and ensure quality.

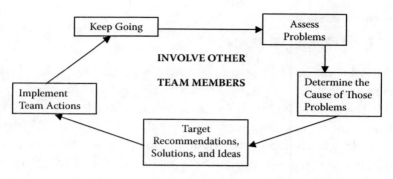

FIGURE 2.2
Team action cycle.

Managers do not control change but manage change.

As teams inside programs and projects become autonomous, they run the risk of pulling in different directions. One team that establishes goals to improve its own processes could subvert the efforts of other teams. When there is a face-to-face meeting as one group, teams are able to agree on proposed planning and project schedules and quality goals or expectations. In meeting as one group, the team will accomplish the following:

- Meet and achieve team objectives
- Resolve conflicts and issues
- Satisfy customer requirements

When struggles with everyday challenges and problems are ignored, a team may use the required team action cycle shown in Figure 2.2.

2.10 TEAM CODE OF CONDUCT

It is okay for a team to fail but to be right at least 80% of the time. Teams that have the privilege and are able to provide clear communication and their own opinions seem to be successful. When one person speaks, listen and treat that person with respect. Once you help each other, you will:

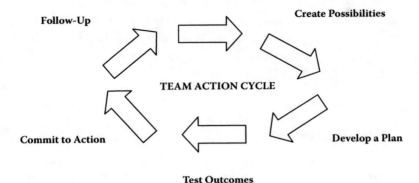

FIGURE 2.3
Team development life cycle.

- Show trust in every individual
- Be honest with your team
- Have ideas that show value
- Stop whining or crying

When teams are expected to attend meetings, be prepared and ensure that action items received are understood in connection with expected goals to be completed. Work together and do not be lazy. Many software designers get themselves in a mode of wanting to be left alone when coding. They get in a zone, so be polite, and do not interrupt, and show respect for other software designers.

The team process includes meetings; promise to honor meeting start and end times. Finally, bring your sense of humor, be friendly and flexible, and always keep a positive attitude. As a software designer, I know the frustrations that could have an impact on jobs and careers in software design/development. Change from an individual to become a team player as shown in Figure 2.3.

2.11 CONCLUSION

Teams should not assume that being knowledgeable would offend others or expect other team members to understand what offends you. The team needs to recognize the relationship between the intent and impacts and stay away from misunderstandings and the scenario of assigning blame. Effective teams need to learn to manage their own reactivity and to be curious about what caused the blame. Practice letting members of a team know how something has an impact on you and rely on others' experience and expertise. There is no "I" in team.

FURTHER READING

Carnegie Mellon, November 2010. CMMI® for Development, Version 1.3, Improving Processes for Developing Better Products and Services. Carnegie Mellon, Pittsburgh, PA.

Cassidy, A., 1998. *CRC Press Practical Guide to Information Systems Strategic Planning*. CRC Press, Boca Raton, FL.

Morris, R.A., 2008. *The Everything Project Management Book*, 2nd ed. Adams Media, Avon, MA.

Pressman, R.S.A., 2010. *Software Engineering, a Practitioner's Approach*. McGraw-Hill, New York.

Wellins, S.R., D. Schaff, and K.K. Shomo, 1994. *Succeeding with Team 101, Tips that Really Work*. Lakewood Books, Minneapolis, MN.

3

Systems Design

3.1 INTRODUCTION

The system/subsystem requirements reviewed by program and project personnel ensure accurate and complete understanding of the restrictions of systems design and applied work products. If program or project plans include reusable software interfaces; requirements are identified and evaluated for use. The term *reusable software* is commonly used in military and aerospace programs or projects. External software interfaces are defined as part of derived software requirements. To support systems design, graphical representations are prepared and take the form of data flow, collaboration/communications, and component diagrams.

3.2 DEFINITION OF SYSTEM DESIGN

The requirements for a system design definition are reviewed with applicable users to ensure an accurate and complete understanding of the restrictions of a system or subsystem that affect work products. The external software interface is defined at those levels and verified for completeness. The program and project plans at times include reusable software and identify interface requirements for use. The external interfaces based on software architecture definitions also are identified as part of derived software requirements.

3.3 SYSTEM ENGINEERING PLAN

The systems engineering team for programs and projects is responsible for the development of software requirements and analyzes the system architecture and design and allocates system requirements. A systems engineering plan (SEP) can be written to establish system-level technical reviews that could be conducted for military and aerospace programs and projects. The major technical reviews and audits affecting software and systems include:

- Initial requirements (IR)
- Incremental design review (IDR)
- Final design meeting (FDM)
- Test readiness (TR)
- First-article inspection (FAI)
- Functional configuration audit (FCA)
- Physical configuration audit (PCA)

The main purpose of the SEP is to address upgraded processes from a systems engineering point of view.

This plan is organized into three main sections: systems engineering, technical program processes, and engineering integration. The systems engineering team describes an orderly and structured approach to the overall system design, software design/development, required formal reviews, and audits. It is important to have such a plan to document and provide the technical expertise to execute activities throughout a software design/development life cycle. Using the plan also enables performance to be more effective and productive and enables technical planners to spend more time planning, ensuring the customer will have greater assurance and satisfaction in addressing the technical challenges that lie ahead.

3.4 SOFTWARE ARCHITECTURE EVALUATION

The purpose of software architecture evaluations is to provide a common approach to developing the work product architecture. This evaluation applies to the implementation of enhancements for change or corrections to existing software architectures. This evaluation provides the feasibility and effectiveness of software architecture definitions to be applied for software work products.

Conflicts in requirements, architecture, or program and project plans should be reported to affected product teams for resolution. The objectives of the software architecture are operational scenarios and system or subsystem requirements. The scope of the software architecture does use interface requirements to analyze operational designs, software risks, and plans to determine the objectives of the architecture.

The development of the software architecture is identified during development and made available and understood before beginning a software design/development life cycle. The program and project plans or schedules are analyzed to determine the impacts on architecture development.

Continual evaluations provide:

- The operational scenarios to be reviewed
- The defined system and subsystem requirements to be analyzed
- The defined system/subsystem interfaces for analysis

Architecture requirements allocate software to gain a complete understanding of the requirements and the capabilities of software architectures. The system or subsystem architecture requirements determine impacts that would include:

- The impacts to quality factors
- The required functional requirements for the determination of the software architecture

The trade-offs between quality performance and the modifications are prioritized and identified outside system or subsystem requirements and reviewed to determine if requirements are to be modified. The evaluation of the software architecture does show how well the architecture meets objectives, constraints, and quality attributes.

The results of software design for architecture changes are examined to determine appropriate design methods to ensure problems are always addressed. One approach to consider is the quantitative technique for the assessment of quality attributes for designs, which are dictated by analysis and considerations and by using your brain.

FURTHER READING

Arlow, J., 2005. *ILA Neustadt, UML 2 and the Unified Process Practical Object-Oriented Analysis and Design.* Addison-Wesley, Boston, MA.

AS9100, 1997. *Aero Space (AS) Standard Quality Management System Requirements— Guidelines for the Application—Part 2.*

Jameson, K., 1994. *Multi-Platform Code Management.* ISA Corporation, O'Reilly & Associates, Philadelphia, PA.

MIL-STD-499B. 1995. *System Approach for Systems Engineering of Defense Systems.* Department of the Air Force STSC Volume 1.

Wigers, K.E., 2003. *Systems Engineering,* 2nd ed., Microsoft Press, Redmond, WA.

4

Software Requirements

4.1 INTRODUCTION

Defined software requirements provide programs and projects with a systematic approach to the development of software requirements provided by various ideas and solutions. Software requirements establish the principals for software design and integration test activities for both software and systems integration. The generation and execution of software requirements are created as a stand-alone item or as an item embedded in higher-level assemblies (i.e., hardware units, workstations, monitor displays, integrated platforms, etc.).

4.2 DEFINITION OF SOFTWARE REQUIREMENTS

Identifying and defining software requirements begin with reviewing the functional or performance requirements developed to identify the constraints on software The system requirement that is allocated to software evaluations determines accuracy, completeness, and applicability of the requirements for work products.

System requirements allocated to software are refined into greater detail to define derived software requirements. Program and project plans that include the capability to acquire reusable software; software requirements are always identified. The tools (i.e., dynamic object-oriented requirements system [DOORS], matrix worksheets, etc.) can be used for the

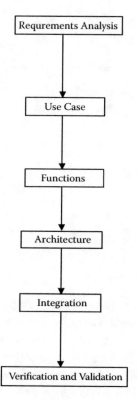

FIGURE 4.1
Software requirements development.

analysis and modeling to gain an understanding of potential architectures and associated software requirements.

The work product for software requirements development is driven by execution and the knowledge of what flows from the start of requirements analysis to verification and validation as shown in Figure 4.1.

4.2.1 Analysis

Requirements analysis includes a step-by-step process to develop requirements for software work products to fulfill high-level user requirements, allocated system requirements, and ideas for system operational concepts. Analysis reports are produced as run procedures are verified and validated to support test and evaluations.

4.2.2 Use Case

A use case is developed to describe a flow of operations for the performance of systems and software implementation. The software use case defines the limitations or technical considerations based on target computers, the execution strategy for a work product, and computer operating systems. Operational cases include functionality, performance, maintenance, and support considerations, as well as the work product's operational environment, including boundaries and constraints.

4.2.3 Functions

The function and architecture definitions are analyzed to ensure an accurate and complete understanding of what software is expected to perform. If program and project plans include reusable software, derived requirements are identified and evaluated for inclusion. Additional knowledge of software operational use cases and the software architecture requires a need to change functionality and associated requirements. This may be an ongoing activity during the software requirements definition process.

4.2.4 Architecture

The architecture interface definition is identified and defined as part of derived software requirements. Graphical representations are prepared and take the form of data flows and software component diagrams.

The functional software design/development life cycle states and modes are established per system requirements. The timing, sequence, conditions, and probability of executing to define and redefine functional interface requirements apply to system architectures.

4.2.5 Integration

The integration of requirements is the transformation of a functional architecture into optimal design solutions. Implementation of disciplined interface management principles is critical for planning resources and to perform systems build integration activities for the execution of meeting software engineering requirements.

4.2.6 Verification and Validation

Software requirements are reviewed to ensure verification and validation. The priority of each requirement is supportive with program and project resources to determine the extent of verification and validation for each defined requirement. The verification and validation for each requirement is identified, and a list of techniques includes analysis, inspections, demonstrations, and tests conducted in software integration facilities.

The most accomplished systems verification and validation of requirements is to plan, evaluate, and record software work product compliance with defined requirements. The risk reduction will assess and ensure software design/development activities satisfy user needs and provide efficient and cost-effective integration of validation and the verification of requirements.

4.3 REQUIREMENTS DOCUMENTATION

The requirements documentation includes software requirements and related information generated, including use cases and derived software requirements, which are the source of each requirement. The software requirements definition is documented according to development plans, software processes, and product standards.

4.3.1 Requirements Traceability

The requirements traceability data are documented according to developing planning, defined processes, and software work product instructions. All software requirements to higher-level requirements or allocated system requirements enter the information into the traceability system according to the program and project's requirements for traceability standards. Software requirements are traceable from system requirements or user requirements and clearly lead to a software architectural component.

4.3.2 Formal Review Preparation

Defined and complete software requirements are critical to have in place before formal review (i.e., functional configuration audit [FCA], physical configuration audit [PCA]) acceptance. Many situations can be discovered during these formal reviews when requirements are not complete and analysis of these requirements is still open or still in work. The FCA is more involved in reviewing requirements that are more related to the release of software documentation and procedures that trace to hardware drawings and are configured in work products (i.e., systems, software, documentation, facilities, etc.).

From my experience, the systems engineering organization or team is more involved in the FCA, which is required to be completed before PCAs are kicked off and conducted. The customer is involved in both audits; once satisfied and the audits are approved, the customer has the deliverable work product in possession.

4.4 MANAGING A REQUIREMENTS TOOL

Senior program and project managers should look for software requirements tools that meet the following:

- Ability to impose requirements in multiple formats
- Support for traceability and impact analysis
- Support for software baselines and releases
- Alerts to modifications of the requirements database

Military and aerospace programs and projects utilize many software requirements tools. A most commonly used tool I am familiar with is DOORS. This type of tool can bring order to chaos; I am not promoting this tool. Any effective software requirements tool allows the capability to manage requirements of the same level so software designers can manage source code. I mention and discuss other software tools pertaining to source code and software design/development in Chapter 6.

4.5 RELEASED SOFTWARE REQUIREMENTS

Various released software requirement definition techniques, organizational capabilities, and process guidance capabilities are designed to address the areas of software design/development that can affect

TABLE 4.1

High Failure Rate for Released Software

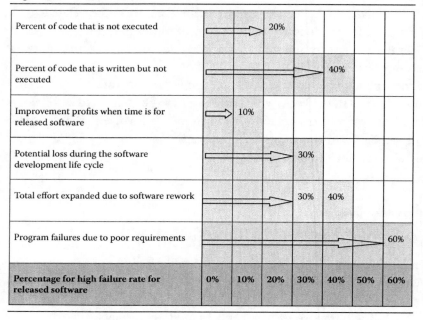

	0%	10%	20%	30%	40%	50%	60%
Percent of code that is not executed			20%				
Percent of code that is written but not executed					40%		
Improvement profits when time is for released software		10%					
Potential loss during the software development life cycle				30%			
Total effort expanded due to software rework				30%	40%		
Program failures due to poor requirements							60%
Percentage for high failure rate for released software	0%	10%	20%	30%	40%	50%	60%

requirements definitions. Directly addressing these areas to align business goals and objectives reduces rework, increases productivity, and ensures that requirements lead directly to program and project success and effective software deliveries to customers. Numerous factors contribute to the high failure rate for released software, but the most significant factor is related to poor and undefined requirement gathering, analysis, and management. The high failure rate for released software is defined in Table 4.1.

FURTHER READING

Ambler, S., 1995. *Software Development, Using Use Cases.* Cambridge University, Cambridge, UK.

AS9100, 1997. *Aero Space (AS) Standard Quality Management System Requirements—Guidelines for the Application—Part 2.*

Gonzales, R., *Requirements Engineering,* 2004. Sandia National Laboratories, http://www.incose.org.

Kazman, R., and A. Eden, January 2003. Defining the terms architecture, design, and implementation, *News at SEI,* 6(1). Software Engineering Institute, Pittsburgh, PA.

5

Software Design

5.1 INTRODUCTION

Software design is a consistent approach and method for the development of software requirements in defined designs of a work product. The software architecture definition provides a framework for the creation of the product design and at times can provide constrictions. The software design definition implements details about a software product's architecture, components, and interfaces. Element traceability of the design and the software requirements is used by software designers. The traceability data and software design definitions are documented according to program and project plans, ideas, processes, and procedures and applicable internal work instructions.

5.2 DEVELOPMENT PLAN

The development plan (DP) for software is a documented and well-defined process useful for implementation and applicable standards. The design model can use a program and project standard, tools, and methods. Software design teams are responsible for supporting the development of software requirements and performing design tasks. The tasks for the development of top-level software design architecture include the identification of major software functions (components), functional hierarchy diagrams, and hardware/software interfaces.

5.3 SOFTWARE DESIGN DECISIONS

The establishment of the software architecture definition provides design concepts and decisions for a work product. The software requirements definition and the software operational concepts identify the capabilities and characteristics required for the inputs that are analyzed and integrated to make key design decisions. Many software design tools, as shown in Table 5.1, benefit the software designer for requirements, code development, configuration management (CM), and software documentation.

5.3.1 Software Requirements Evaluation

The reviews and evaluations of software requirements define that software operational scenarios ensure problems affecting software design

TABLE 5.1

Software Design Tools

Software Tools	Description
Requirements analysis and design tools	Requirements analysis tools will be used by software development organizations for requirements analysis of new software. Organizations, which do not use the program-wide standard, provide requirement documents for inclusion in the program database. Commercial off-the-shelf (COTS) tools can be used for software design/development and documentation to be used to document reused software categories. Requirements and design documentation retain the format of the tools.
Code development tools	Code development tools for software are proven in the design/development of the product line or work product software. The tools, such as code editors and compilers, are employed.
Configuration management (CM) tools	CM tools supports distribution of incremental development processes implemented in software companies and for military and aerospace program and projects.
Commercial off-the-shelf (COTS) documentation tools	COTS tools include standard word-processing and graphic development tools to provide for the development and maintenance of documentation with the delivered software.

are identified, evaluated, and resolved. The software design/development team performs a risk analysis using prototype software to help support early requirements evaluations and design feasibility. Information from these evaluations is fed back into the output of the software requirements development phase if the requirement is proven to be unusable and not to be implemented for use.

5.3.2 Software Reuse

The reuse of software components identifies evaluations by software architecture definitions on how to decide on the incorporation of components into the software design. Opportunities for software reuse support numerous software product developments in state and international markets. The reuse criteria are identified in defined software plans to determine if the program and project reuse library or existing software work products can be used for near-term software design activities.

5.4 PEER REVIEWS

Software processes require design engineers to conduct and perform peer reviews to find and correct as many errors as possible before test team integration or customers find problems during delivery. The peer review starts with requirements, design models, and uninterrupted code and unit tests for the software designer. These reviews are applied at various stages during the software design/development life cycle to create clean software work products and provide the assurance that issues or errors are discovered and resolved.

If errors are found early in the development life cycle, time is saved, and the cost is not a concern. Minor software errors that are not fixed or resolved become major errors later in a program and projects. Software design engineers always make mistakes in their code development, so early peer reviews reduce the amount of rework and are not required late in the program and project.

As stated in CMMI® for Development (version 1.3) (CMMI stands for Capability Maturity Model Integration), peer reviews are an important part of verification and a proven mechanism for effective defect removal. An important corollary is to develop a better understanding of the work products and the processes that produced them so defects can be prevented and process improvement opportunities identified. Peer reviews involve a methodical examination of work products by the producers' peers to identify defects and other changes that are needed.

Examples of peer review methods include the following:

- Inspections
- Structured walk-throughs
- Deliberate refactoring
- Pair programming

The peer review verification methods identify software bugs, errors, and defects for removal with recommendations to improve code development as shown in Figure 5.1.

The peer review method is applied to software work products developed by programs and projects, but it can also be applied to other work products, such as documentation and training, which are typically developed by other software teams. Preparation for peer reviews includes identifying affected teams or groups to participate in the peer review of affected software work products. The criteria for conducting peer reviews are as follows:

- Schedule the peer review at a convenient time
- Assign reviewers (i.e., teams)
- Prepare or update materials
- Provide peer review checklists
- Introduce training materials
- Select software work products
- Provide entry and exit criteria (i.e., minutes, action items, etc.)

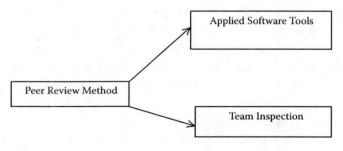

FIGURE 5.1
Peer review method.

To ensure you have a successful peer review, make sure you have selected the right reviewers to be involved and guidelines are understood from the start. If peer reviews are conducted and performed correctly, the peer review was performed and done right.

5.5 SOFTWARE DESIGN/DEVELOPMENT SUGGESTIONS

I suggest we look at two software design/development methods. One method is concurrent software design/development, which is a technique to reduce the time to improve productivity through the simultaneous performance of activities and processing of information. The concurrent software design/development method refers to tasks that are performed simultaneously by different teams or groups that support a team approach to development. The second method is Lean software design/development. According to this method, it is far more effective to have small working teams across the boundaries of informational handoffs, reduce paperwork loads, and maximize strong communication.

5.5.1 Concurrent Software/Design Development

Concurrent software design/development activities require software designers who have enough expertise to anticipate where the defined design is going. When starting software design/development, only partial requirements are known and developed in short iterations to provide feedback for systems to emerge. The use of the concurrent software design/development method does make it possible to delay commitment until the last moment when failure to make a decision eliminates an important alternative or decision.

5.5.2 Lean Software Design/Development

The Lean software design/development objective is to move as many changes as possible from the top curve to the bottom curve as shown in Figure 5.2.

Lean software design/development delays the freezing of all design decisions as long as possible because it is easier to change a decision that has not been made. This type of software design/development emphasizes

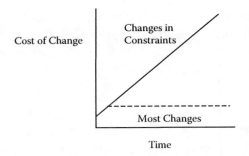

FIGURE 5.2
Cost curves.

designing and managing changes throughout the life cycle. Better understanding of software engineering and quick delivery to customers benefits the concepts to improve processes and increases quality according to the following principals:

- Early software product development
- Elimination of wasted time
- Understanding and working to software requirements
- Meeting customer expectations and deliveries on time
- Achieving and implementing team goals
- Shortened design and test software life cycles

5.5.3 Lean Software Configuration Management

The traditional software configuration management (SCM) practice involves the identification of systems and software design/development and providing configuration control. Selected work products and the descriptions to maintain traceability of those configurations are key points throughout the software life cycle. The Lean concept is the process to compare common information with Agile software development.

5.6 AGILE SOFTWARE PROCESSES

The implementation of Agile software processes, principles, practices, and software design/development deliveries of work products to customers

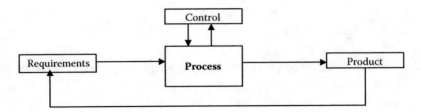

FIGURE 5.3
Agile management model.

does provide fewer defects. Application of Agile methodologies supports numerous initiatives and provides a program and project with a manager's approach to emphasize short-term program and project planning. The adaptability to changing requirements as well as close collaboration with customers and affected teams show accountability. The Agile management model consistently depicts processes as shown in Figure 5.3.

The Agile model adopts values that are consistently making decisions that may cause a rejection of a software design. Agile models are more effective than the traditional (i.e., waterfall, spiral, etc.) models due to perfection versus good-enough concepts for software design practices. The software design engineer using Agile concepts has the capability to understand information first before jumping into software design/development activities.

The current state of the economy changes each day. We must resolve the software engineering approach to adapt to Lean processes and meet the needs of programs and projects. The four key elements for Agile software engineering are:

- The team has control of work assignments
- Communication with team members and customers is needed
- Change is good: "Think outside the box"
- Customer satisfaction and expectations are achieved

The Agile process method for team efforts reflects how a team of software people work together. An Agile process continually improves processes that are not working or are causing major delays in the software design/development environment. Internal program and project managers try to keep the team together by allowing decisions, expectations, and a commitment to show results. When the Agile team working its own processes at times does discover problems, the team will stay the course to solve problems that could have an impact on these processes.

The Agile method is also about continuous incremental delivery of products such as software and systems to other program and project team members and the customer. The Agile team explores and evaluates work product needs and requirements. The planning and analyzing of what to build and defining acceptance are an advantage of testing software and coordinating efforts that feed from one team member to another. Whichever Agile or Lean framework, method, or techniques are used, they employ such things as:

- Data models
- Rules of engagement
- Guides or maps
- Agile team rules

These items can be helpful as teams explore the designs and builds that are prepared for software and systems integration tests to be conducted and performed.

Agile provides team interactions that deal with processes and tools. Performance through team members boosts accountability for results and shared responsibilities for team effectiveness. Strategies, processes, and practices improve effectiveness and reliability. A successful Agile team stays alert to change and will adjust strategies and practices to match.

5.7 CONFIGURATION MANAGEMENT

When we look at the Agile process, CM methods are not referenced for any specific routines. These methods are a supporting discipline and not directly involved in creating executable code. If Agile processes have a lack of configuration control, then Lean principles are a bust or a big waste of time and lead to a chaotic activity. You see no progress in software design/development.

The members who make up the Agile team are focused on processes and tools that would imply configuration control is not important, but CM disciplines aim to trim the process and provide more automation in the tools, bringing back focus to configuration control objectives. The software tools common to other team members are adapted to the processes.

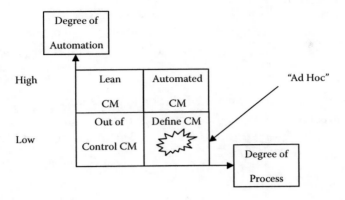

FIGURE 5.4
Lean CM performance.

CM is a support discipline to the Agile team, but there are times in software design/development whether work products are comprehensive to plan and process documentation. Controlling change is the foundation to ensure software design/development activities are important when change is in the picture. Make sure CM methods do not limit change and become a stumbling block and a nuisance for program and project plans.

By adapting Agile practices in the CM process, teams can have a leaner concept in programs and projects by:

- Ensuring change is common with configuration control objectives
- Having a clear distinction for changes to requirements
- Embracing that change is possible with appropriate routines for configuration activities
- Providing the development and enhancement of Agile ideas

My experience has been that common changes from Lean CM create chaos in working toward heavy and out-of-control processes. An example is defined in Figure 5.4.

5.8 SOFTWARE STANDARDS

It is and will always be a requirement that software design engineers follow required software standards to ensure that development processes are in accordance with specific process models (i.e., CMMI). The software design engineers are required to show defined, managed, and consistent improvements during life-cycle development. At minimum, software standards consist of:

- Documented and maintained plans and procedures
- Peer reviews to eliminate defects and prevent future occurrences
- Standard software tools for design, code and unit test, and configuration control

5.9 CAPABILITY MATURITY MODEL INTEGRATION

CMMI provides the best opportunity to address software design or development with ongoing support to customers after delivery. The design/development process could be a complex activity that benefits the accomplishment of correct software tasks during the life cycle. Software engineering and processes require an association with each other when it comes to software engineering. CMMI does provide a systematic, disciplined approach to all software engineering tasks in the affected program and projects. The development of software work products using CMMI will enhance the knowledge base for software designers.

To implement CMMI processes provide the content for performance during the software life cycle for change, development, installation, integration, and maintenance. Complete software design practices provide the basic concepts (i.e., form, fit, function, interface, integration process, etc.) and other sound concepts. Some software engineering tasks defined by CMMI are:

- The identification of internal and external interfaces
- Software design to establish infrastructure abilities during software design/development
- Development of plans, processes, and procedures
- Reuse of capabilities for software identified for use

With CMMI implemented and used by the program and projects, this model represents processes in two different scenarios: continuous and staged. Every organization should work toward the achievement of a third level: defined. The process standards apply to:

- Requirements development
- Technical solution (see following discussion)
- Product integration
- Verification
- Validation
- Organizational process focus
- Organizational process definition
- Organizational training
- Integrated project management
- Integrated supplier management
- Risk management
- Decision analysis and resolution
- Organizational environment for integration
- Integrated teaming

Software engineering methods have the capability to utilize many tasks and activities along with described approaches. The method is effective when how much support is needed. The engineering process area of technical solution can be the template for software engineering to design, develop, and implement solutions to requirements supporting software and systems integration. The processes that are related to this process area concept receive requirements for managed processes.

Technical solution processes support each other and service products related to the software life cycle. A technical data package provides the software design/development team complete understanding for effective design and development for the next phase of integration activities.

5.9.1 CMMI Version 1.3

In 2009, CMMI version 1.3 was initiated. The release was to provide a new approach to software companies and military and aerospace programs and projects to improve performance in appraisals and training. The focus is on:

- High maturity
- More effective processes
- Conducting and performing effective appraisals
- Commonality in all product suites

Major elements implemented allow appraisal teams to become more effective in reflecting organizational high maturity using a staged approach.

The high-maturity practices were not understood and were unclear, leading to mixed views by organizations for how objectives are related and lead to high levels. Modernized practices include improvements in:

- Agile environments
- Functional requirements during software design/development
- Subcontractor agreements pertaining to COTS (commercial off-the-shelf) and NDI (non-development item) software
- Organizational training

CMMI version 1.3 coverage will add updated information currently supported by Lean and Six Sigma and customer satisfaction for software design/development life-cycle tasks.

5.9.2 Lean Six Sigma

Lean and Six Sigma tools and philosophies have helped thousands of software companies and military and aerospace programs and projects dramatically improve processes, customer satisfaction, on-time delivery, and other measurable results. But, do these same tool sets apply to the processes of software design/development? The answer is "yes" when the correct tools are applied in the right way and to the right process.

The Six Sigma methodology attempts to reduce process variation, resulting in fewer errors and defects. A software defect is defined by customer requirements, whether formally documented or an expectation that is not met. A defect may be detected during the software design/development phase by team members or later when the customer is using the delivered software. The Six Sigma process does show fewer defects per opportunities or zero defects in a software work product.

Opportunities for defects abound, including but not limited to macro-functional requirements allowing the end user to enter wrong data.

To accomplish the goal of zero defects, team members must have highly structured and robust processes for each step in a software life cycle. In Six Sigma, the steps are:

- Define
- Measure
- Analyze
- Improve

The control software teams often use a form of these requirements: gathering, design, implementation, verification, and maintenance. The formal processes program and projects' scope are customer requirements to have effective methods for software and systems integration. Data flow analysis and feature breakdown structures ensure fewer opportunities for errors. During the software design/development process, the Six Sigma philosophy is applied for building quality through mistake-proofing methods. The creation of effective charts of when and where defects were detected and code had to be rewritten, added, or reused can assist the team in evaluating which steps in the process have the most variation and are candidates for Six Sigma process improvement.

The Lean production of software and systems integration work products focuses on the elimination of waste from defined processes. The eight wastes are easily remembered with the acronym DOWNTIME:

- Defects
- Overproduction
- Waiting
- Nonutilized talent
- Transportation
- Inventory
- Motion
- Excess processing

Software design/development is a complex process integrated with wastes that include defects as discussed and resulting in rework or reuse, which is another waste for excess processing. Waiting on waste occurs when programmers, project leaders, and team members require information such as customer requirements and parameters from code and unit tests that could delay their development of software work products. Excess processing of waste also occurs with numerous review cycles rather than having a robust process for designing quality and being right the first time. An example of overproduction and the inventory wastes may be the creation of features that were not requested or are not needed.

The Lean philosophy of problem solving uses simple and straightforward tools to achieve fast yet powerful results. Utilizing the software design/development process creates a process flow or a value stream map that shows each step. Look for each of the eight wastes as you walk through the process. Identify the "hidden factory" or tasks that are regularly performed but are not documented or do not add value. If an activity does not change the functionality of code or the software programming activity, it is waste. If the activity does require waste, such as some phase quality gate code reviews, do minimize the wasteful activity by creating a sound process for performing the task and minimizing rework or reuse.

When you have identified waste in your process flow, drill down to the root cause using a simple technique. Develop a future state process that eliminates or minimizes waste. Finally, implement process changes, putting in place standardized work products and program and project leaders to follow up and ensure improved processes function the intended way. Make progress so the achievement of program and project milestones is visible to ensure team members can see progress and are accountable for meeting their goals. Do not forget to track waste by moving forward so you can continuously improve processes. A schedule showing the number of times a section of code has to be rewritten or reused is an example of how easy it is to identify, track, and eliminate waste.

Combining the data-driven approach of Six Sigma and the waste-eliminating tools of Lean streamlines the design/development process and produces better software in less time. The goal is to satisfy your customers with amazing work products and services. Creating a defect- and waste-free process during the software life cycle does make excellent programs and projects and the customer happy.

5.10 SOFTWARE COMPANIES

Many software companies and military and aerospace programs and projects previously and currently build the concept of specializing in software design/development to boost competitiveness in the software industry. Many global companies have always gained ground to integrate the market with strong software and hardware. Countries have been in competition in software development for years, nurturing technicians with software expertise at an early stage.

5.10.1 Software Design/Development

The software design/development opportunities secure databases related to software technology owned by companies, governmental research facilities, and even universities to provide technology concepts to customers. The software technology consultant provides customized consulting and technology to companies, government, and individual and small firms.

The lower limits of software integration restrict the consultant from participating. The software companies have increased the dollars for winning business and sales. Effective methods for software and systems integration efforts inject profit from inside the software design/development sector. Profits will increase once actions for consulting come into play for software companies all over the world.

5.11 CONCLUSION

The pairs of software design/development attributes shown cannot be exhibited simultaneously without circulating the brain. One can (and must) learn to switch, change, and be flexible from one mode to the another as needs arise. This can be done, and one can learn how to do it.

The attributes of a good software designer/developer are the following:

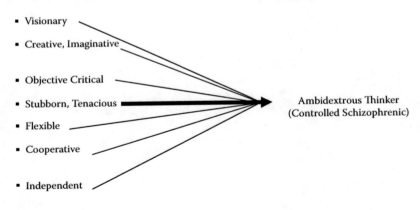

- Visionary
- Creative, Imaginative
- Objective Critical
- Stubborn, Tenacious
- Flexible
- Cooperative
- Independent

Ambidextrous Thinker
(Controlled Schizophrenic)

Yearn for the unachievable.

FURTHER READING

Blackburn, J.D., G. Hoedemaker, and L.N. Van Wassenhove, 1996. Concurrent software engineering: prospects and pitfalls. *IEEE Transactions on Engineering Management*, 43, 179–188.

Cantor, M., 2001. *Software Leadership: A Guide to Successful Software Development*. Addison-Wesley, Boston, MA.

Carnegie Mellon, November 2010. CMMI® for Development, Version 1.3, Improving Processes for Developing Better Products and Services. Carnegie Mellon, Pittsburgh, PA.

Cedro, T., 2011. Master Black Belt, PMP, MBA Lean Six Sigma Toolbox.

Chaplin, C.R., 1989. *Creativity in Engineering Design*, United Kingdom Fellowship of Engineering.

Getting a handle on process, 2010. *CrossTalk: Journal of Defense Software Engineering*, 23(1).

Humphrey, W., 2006. Sweet Predictability. *World of Software Development*, 14(2), 14–17.

McNaughton, A., 2004. *Software Development*. CMP Technical Insight, LLC. Raleigh-Durham, NC.

Peters, L., 2008. *Getting Results from Software Development Teams*. Microsoft Press, Redmond, WA.

Poppendieck, M., 2002. *Lean Software Development*. Addison-Wesley, Boston, August 2003, Vol. 11, No. 8.

Schwaber, K., and M. Beedle, 2001. *Agile Software Development with SCRUM*. Prentice-Hall, Englewood Cliffs, NJ.

Toro, C., 2011. *Lean and Six Sigma Toolbox*, Master Blackbelt, PMP, Meridian, ID.

6

Software Implementation

6.1 INTRODUCTION

The software implementation method provides assurance that software engineering builds function as expected in target software and systems environments and enables smooth execution for verification and validation activities. Disciplined software implementation principles, planning, and resources for systems buildup provide effective testing to be conducted in a development facility for a software/system integration environment. Software released under configuration management control is described in a defined documented configuration management plan (CMP) to provide the necessary requirements for software implementation inside integration facilities.

6.2 CONFIGURATION MANAGEMENT

The configuration management software team or organization ensures that configuration management practices are applied consistently throughout the software life cycle for work products that are developed and maintained by programs and projects. The team focuses on identifying and managing changes and maintaining software configuration and documentation visibility.

The configuration management concept is a cross-functional process applied over the life cycle of a software work product and provides visibility and control over functional and physical attributes. The processes that are used during all phases of software design/development provide the necessary disciplines that identify applicable products, establish and control baselines, and document and track changes to those baselines. Also, configuration management processes control the storage, access, changes, archive, and release of the software work products.

This team develops operating procedures that describe implementation of processes required to satisfy the requirements and direction provided under associated and documented plans.

6.2.1 Build Requests

When software engineering builds are requested, electronic files or hard copy paperwork is written to provide build checklists to assemble, compile, link source code, build archive copies, and provide listings for use in software design/development, test, and work product deliveries to customers.

Automatic generation of build deployments ensures customer confidence in the releases. For the program and projects to be successful, processes used by build engineers include the capability to package builds and documentation together. Creating an approach to meet build and installation processes requires coordination between internal and external teams to become efficient and available when supporting scheduled tests or configuration checkouts.

The build engineer with the direction or authorization for a requested engineering build has a defined role to perform tasks related to software construction and configuration control, including the following:

- Creates build folders to store documentation of software building
- Provides source code changes and control of the source code
- Maintains and controls records during program and project development

6.3 CONFIGURATION MANAGEMENT TOOLS

The management and use of effective configuration management software tools provide version control and change management concepts. The tools (i.e., ClearCase and ClearQuest) may be used to provide the capabilities

TABLE 6.1

Configuration Management Tools

Tool or Vendor	Software Activity Support	Host System	Purpose of SCM Tool
ClearCase (IBM)	Design, code, and unit test, software builds/installation, integration, and test	UNIX/PC	Tools for documentation and source code, support multiple developments, and release baselines
ClearQuest (IBM)	Code and unit test, integration, and integration testing	PC	Software problem reporting, logs, tracking, and software debugging and fixes
FORTE (SUN) APEX Rational Clearmake (IBM)	Software engineering builds	UNIX	Compile and build released software executable products
Microsoft Office	Software engineering activities	PC	Support documentation, software design/ development, e-mail communication, and data analysis

for adding new files to a software design/development environment and provide version control to applicable directories and files. File sharing, parallel software design/development, multiple team support, and software reuse are essential for meeting integration test activities demanded by the schedule. The configuration management software team administrates or manages software tools. Table 6.1 provides an example and overview of tools included in a software and systems integration environment.

6.3.1 IBM Rational ClearCase

The configuration management tool for software design/development that I have experience with and used in military and aerospace programs and projects has been IBM Rational ClearCase. This software tool is an object-oriented database utility provided to establish software product archiving, automation, identification, version/change control, engineering building, product releases, status accounting, and auditing activities. The ClearCase software tool provides an open architecture to implement configuration management and control solutions. Web site content for computer software companies, the military, and aerospace industries employs many different development environments.

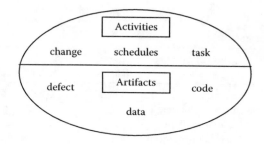

FIGURE 6.1
Unified change management definition.

The concept of unified change management (UCM) combines activities and artifacts as shown in Figure 6.1.

Many programs and projects use different names for ClearCase roles and assignments, but configuration management can use the following as an example:

- Architect: Understands and sets up the system architecture
- Configuration manager: Familiar with change and control processes
- Lead: Responsible for schedules and assignments
- Software design engineer: Makes changes to files under configuration control
- Build engineer: Utilizes software build concepts and tools

A functional overview of a ClearCase concept is the repository named version object base (VOB). This is a data repository where files, directories, and data are stored. All files and directories are managed inside the VOB and can expand from hundreds of files and directories to thousands. Table 6.2 defines the ClearCase—UCM roles, responsibilities, and main

TABLE 6.2

ClearCase—UCM Roles and Responsibilities

Role	Main Objectives
Architect	Define models (architecture)
Configuration manager	Set up configuration management environment (i.e., repositories, importing files, etc.)
SCM lead	Assign and schedule work activities and define written software configuration management policies
Software design/developer	Make changes to files/directories and deliver software to build engineers
Build engineer	Builds components for established baselines ready for test

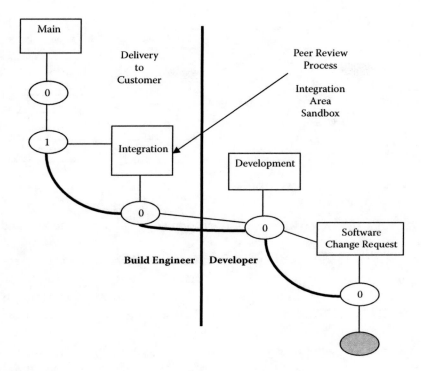

FIGURE 6.2
ClearCase VOB architecture.

objectives. The files and directories can be moved or transferred to other VOBs when the repository becomes too large. They can also be split and work together. The ClearCase architecture and the VOB database ensure the checkout of files and support data recovery if needed.

The ClearCase VOB structure example is shown in Figure 6.2.

6.3.2 IBM Rational ClearQuest

The change request management process is critical when reporting any requests from team members that are needed to change or update software and systems integration work products. IBM Rational ClearCase comes into play with another software configuration management (SCM) tool, IBM Rational ClearQuest.

This software tool provides support for change request management processes and is a complementary tool for ClearCase. The database utility is used for recording, tracking, and reporting and provides internal access control mechanisms for permitting the restriction of work product updates

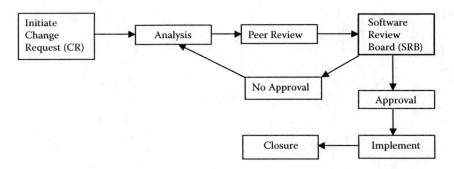

FIGURE 6.3
Change request process.

at various stages of software design/development, integration and test, and production processes.

The change request is a request from team members to change an artifact or process. Documented in this type of request is the information for problems occurring during software design/development and impacts that could occur. An example of the change request process flow from initiation to closure is shown in Figure 6.3.

The change request administrator maintains a tracking system to control software code and software documentation change status. The administrator manages the database, coordinates, and provides software inputs to program and project changes to establish traceability to a higher-level change authority with an impact on software. Software plans and process procedures provide the information to coordinate the required review boards to maintain records, including status of change requests, reports, and documented releases.

This review board is established for software teams to review and disposition changes that affect controlled software and related documentation. All software changes are documented, approved, and implemented per a change request. The review board meetings are scheduled and coordinated by the program or project manager, leader, or designated representative serving as the review board chairperson.

The review board members include, as a minimum, those from the following areas:

- The affected software teams
- Configuration management
- Test
- System engineering

- Quality
- Security (if change has an impact on classified or trusted software)
- Change sponsor

The major activities performed by the review board are evaluations and dispositions of change requests, assignment of priorities, review of action items, change dispositions from prior meetings, and the evaluation of deviations that occurred for discussion.

There are many configuration management software tools used in military and aerospace programs and projects that can be discussed. The software tools selected are required to fit the environment used by the teams and for software and systems integration activities. Do not take my word that the Rational ClearCase and ClearQuest are the only usable software tools. Other tools will work and support the program and projects, so they will be okay.

6.4 SOFTWARE MEDIA AND DATA

The physical software media (i.e., disk units, CD, DVD, hard drives, etc.) identification and media labels must also be in accordance with the program and project documented media requirements in affected plans.

Marking information could be displayed electronically on the exterior of the physical media containing the software or provided within the media through a file in each piece of software data or a written set of electronically submitted files (identified as .doc, .txt, executable files, etc.). These media files reside in a computer media library (CML) for engineering use.

Copies of the software media generated are verified and validated by a quality team.

An example of what could be documented on a media label follows:

Date: Day/month/year format
Title: Document the title of the software being produced

- Derived from: Program and project
- Special handling: Distribution requirements
- Contract number: Document contract number
- Part number: Document software identifier
- Software version: Media version

6.5 FUTURE TRENDS

There are major improvements in software technology and future trends for effective use of software tools. With the technology, there will be time to address and resolve issues and improvements required for:

- Software design/development
- Software process definition and enhancements
- Reuse of software program and project artifacts
- Ongoing support of past tool artifacts
- Training for software design engineers
- Software tool disciplines

The software design/development emphasizes SEE (software engineering environment) technology to allow detailed definitions of the required roles and responsibilities for users and if organizations related to the program and projects are ready. Acquiring software tools should not be a solution to show tools that are out of control for software design/development but ensures processes are defined for the management for software development activities.

Having effective software tools in place will improve software design/development and quality produced and increase the productivity for software and test engineers. The insertion of SEE technology in the program and projects is successful when implementation plans are well defined.

The main reason for software tools and how they are adapted to requirements should be based on how these tools can approach implementation of the design/development. In the future, many tools for software design/development will support the life-cycle work products and the processes defined by the user. The major obstacle will be the balance and control for a stable software implementation plan and for adapting to changes that occur.

6.5.1 Tool Support

When a program or project is ready for software tools that will be effective during design/development, the key is selecting the right vendor products to match engineering needs. Questions are asked, and the primary steps for organizational needs are as follows:

- Become effective for designing and developing work products
- Establish the resources for use of software tools
- Conduct software implementation with no problems
- Conduct training

6.6 CONCLUSION

The major building block of software design/development improvement is to make sure the automation of software tools is understood. Costs are significant for short- and long-term use. In programs and projects, it is critical that the organizations enhance software implementation for productivity and quality.

FURTHER READING

Gustavsson, A., October 1989. Maintaining the evolution of software objects in an integrated environment. In *Proceedings of the Second International Workshop on Software Configuration Management*, ed. Richard N. Taylor, 117–117. ACM, New York, doi: http://dx.doi.org/10.1145/72910.73355.

Fayad, M.E. 1996. IEEE Computer Society. *Controlling Software Development, MIL-STD-973, Configuration Management*, Reno, NV.

Hanrahan, R. 1994. IEEE STD 109-1994. *IEEE Recommended Practice for the Evaluation and Selection of CASE Tools*, STSC Hill Air Force Base, Clearfield, UT.

Herrmann, D.S., 2000. *Software Safety and Reliability*. Wiley-IEEE Computer Society Press, New York.

Keyes, J., 2004. *Software Configuration Management*. CRC Press, Boca Raton, FL.

White, B., 2000. *Software Configuration Management Strategies and Rational ClearCase®: A Practical Introduction*. Addison-Wesley, Upper Saddle River, NJ.

7

Software Integration

7.1 INTRODUCTION

The methods for software integration provide required steps to be conducted for integration and checkout of informal software engineering builds. The software design/development team and test engineers need to develop a strategy for planning, design, execution, data collection, and test evaluation. The software integration activities are informal and flexible for software checkout to prepare for the software and systems integration phase of the work product.

7.2 SOFTWARE INTEGRATION STRATEGY

The strategy for software integration provides a road map that describes the steps to be conducted as part of the implementation of software to start integration activities. When a strategy is planned, resources are required. This strategy should be flexible and promote an approach that could show change. Planning by senior, program, and project managers needs to track program and project progress and will require the following characteristics:

- Effective technical reviews should be conducted
- Different integration techniques and software approaches are shown
- Software designers are required to be involved from the start to the finish

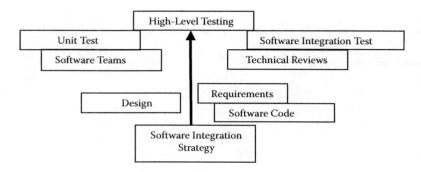

FIGURE 7.1
Software integration strategy.

The software integration strategy provides an example of higher-level integrations (Figure 7.1).

7.2.1 Approach to Software Integration

The approach to software integration activities is planned in advance and is the start for effective software integration. This approach accommodates lower-level integration to verify software code development that has been implemented correctly and validate major system functional expectations by customers.

The approach of effective planning for software integration provides guidance software design/development and test teams to reach milestone expectations of senior, program, and project managers. The steps for effective software integration occur numerous times as deadlines occur, and measurement problems are resolved early in schedules.

7.2.2 Software Integration Testing

What is software integration testing? The concept for testing software is to uncover errors, troubleshoot, and fix problems that occur during a test. Test plans and procedures are developed to test systems and, if required, rerun integration tests that are to be witnessed by quality assessors or customers.

The software test plans or procedures developed by program and project managers along with testing experts ensure that testing strategies are not wasted time during integration. Errors can appear that were previously undetected. That is the purpose of having plans and procedures in place.

Test specifications are also defined and documented to provide testing steps that test conductors or that experts can implement.

Performing a review of test specifications prior to software integration testing is a strong attribute assessment before tests are complete. An effective approach to utilize a test plan or procedure for software leads to the order and discovery of errors at each stage in the test integration process.

The techniques for developing and construction of the software architecture goals take unit-tested components and build program structures established by design. The "bam theory" approach is to attempt nonscheduled software integration and testing. This approach is performed in the following three steps:

- Software test plans, procedures, or internal work instructions are ready to support integration
- Software integration is ready for testing to be conducted and performed by all notified team members
- Control must be maintained between multiple tests running at the same time. Lack of control can cause chaos

7.2.3 The Big Picture

Software processes are viewed as a spiral concept (Figure 7.2) for software integration to ensure testing is implemented for software design/development.

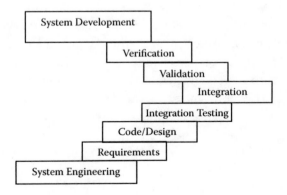

FIGURE 7.2
Spiral concept.

7.3 DEVELOPMENT FACILITY

Early in the software design/development phases for military and aerospace programs and projects, a Development Facility (DF) is normally established for software integration activities. This facility is used for the preparation of software prior to delivery to a software systems integration facility (S/SIF). Many statements or comments are made about these facilities and whether they have an effective way to test traffic loads on specific work products. In discussions with technicians and test teams, I have tried to have in place effective methods for software systems integration testing in order to show that we need traffic load tests in these development facilities.

An overview of developer facilities includes geographic locations where software integration is performed, facilities used, and secure areas along with other features. Customer-furnished equipment, software, services, documentation, data, and facilities are required according to contractual agreements along with a schedule detailing when these needed items are included. Other required resources include plans for obtaining the resources, dates needed, and availability of each resource item.

The engineering design/development teams are primarily located in a designated software development geographic location.

7.3.1 Software Operations

There is adaptation intrinsic to software operations. Examples of this include parameter-based initialization data and settings selected or entered by a software designer/developer and test teams during operations of the software and systems retained for other test integration purposes.

The requirements for a software design/development environment must be understood when a schedule calls for software development and integration activities to be performed. Software integration plans ensures that each element of an environment performs to intended functions in support of the software design/development activities. The plans also provide requirements for test environments to perform software testing, including integration, troubleshooting, and checkout to ensure that each element of the test environment performs intended functions.

Software applications or tools used for designing, building, or integration testing the work product could be deliverable. Any nondeliverable software on which the operation depends can be identified after delivery and provisions made to ensure program and project sponsors or stake-holders obtain the same software and work product.

Software tools used for integration and hardware units installed are placed under configuration control. When software upgrades or new versions become available, program and projects evaluate and recommend whether the updates should be incorporated. Upgrades are installed as soon as is reasonable for the design/development, and integration activities are agreed to by all affected organizations. The criteria for evaluating an upgrade include considering integration problems detected, problems solved, and impacts on software integration efforts.

7.3.2 Software Configuration

All software configuration identifications documented in accordance with the program or project software plans are effective ways to ensure configuration control. The configured baselines identify the development life cycle, namely, functional and allocated work product baselines. Unique software documentation and media define software configuration baselines.

7.4 SOFTWARE INTEGRATION SETUP

The software integration setup method involves planning with program and project managers to coordinate with the facility operations manager. Allocated resources such as computers (i.e., workstations) and hardware units are provided to the software designer/developer and test teams to conduct informal integration testing. The software engineering builds and loading into hardware units are performed by selected build engineers.

7.4.1 Integration Test

Inside programs and project integration facilities, system integration tests are conducted. Verification steps ensure tests provide a check of the capabilities of software and hardware units. The software integration test is to be repeated numerous times and ensures all integration test problems are resolved and performance is accomplished early in the defined system and the system is working to software requirements.

7.4.2 Installation Plans and Procedures

The installation plans and procedures define systems' specification requirements. The plan and procedure for the software integration tests cover the testing of requirements and verification methods conducted in the DF. Specific integration test plans and procedures consist of checkout activities to ensure system utilization. The integration testing environment provides necessary steps to be followed, data collected, and analysis solutions are used or implemented to produce test reports at the end of testing activities. The installation test plans and procedures are to be peer reviewed and approved for release by program and project managers to prepare for the start of software integration testing.

7.4.3 Integration and Checkouts

Early integration and checkouts focus on software components applied to tests to uncover errors. Once the components are tested, an informal system is constructed. Tests are executed to fix software bugs and errors. The recommendation of processing "draft-only" test plans and procedures provides loading instructions, execution, and the capabilities for uncovering problems early during integration testing. The software design and test engineers need to troubleshoot as early as possible before going into a formal test environment.

7.5 SOFTWARE INTEGRATION LOG

A software integration log provides a view of the day-to-day operations for the design and test teams using hardware units for integration and checkout. Facility operations managers use these logs to support operational setup activities. There are no formal released plans or procedures required during this informal phase of integration. Quality personnel are not required to support integration and checkouts performed by the software design and test teams. This is an effective method for conducting informal software testing in preparation for such activities in a facility for software and systems integration. Allow software design teams the freedom to fix and debug problems and work with test teams to ensure plans and procedures will be ready for release to support formal test phases.

This software integration log is an effective method for the software design and test team to troubleshoot problems discovered during this informal test phase. Once the software is loaded into hardware units, the software does not have an impact on and take hardware out of configuration. I repeat: Software loaded in hardware units *does not* and *will not* have an impact on and take hardware out of configuration. Hardware quality and quality software teams butt heads concerning this issue. The quality team for software provides the correct answer, so please hardware quality follow its lead. No formal software plans such as step-by-step operational paperwork or tools in the manufacturing environment are required.

7.6 SOFTWARE TEST COMPLETION

The term *acceptance testing* is discussed throughout a software integration program and projects. There are always questions when this topic is mentioned or discussed. When will the software integration testing be completed? There is never an accurate answer, and that frustrates program and project managers. The burden is always on software engineering. Remember that the importance of quality is first and not second in any software program and projects. The pressure is on when integration testing keeps going on and on and not completed in time to deliver the S/SIL work products to the customer.

The metrics collected or testing models make it possible to develop guidelines for many answers to the question of when software integration and testing will be completed. Software integration is the first phase before any stage of systems integration. Understand that metrics do come into play in the early stage of software integration and testing. All program and project managers need to implement and use metrics instead of solving problems with no data to support and make key decisions.

7.7 INTEGRATION VERIFICATION AND VALIDATION

One of the important software processes for integration is the element that is often referred to as verification and validation. The verification aspect is a set of tasks that ensure correct implementation techniques are in place to verify that the right work product is being integrated correctly. The validation concept ensures that the correct work product is the right product to validate. The quality team roles are to perform:

- Technical reviews
- Configuration management audits
- Progression monitoring during software integration
- Plans, procedures, and documentation reviews
- Qualification and acceptance testing
- Witnessing of implemented plans and procedures during integration and testing

Quality during software integration throughout the life cycle shows that proper methods and tools are being utilized. The real motive for quality can be applied for very large- and small-scale systems.

7.8 CONFIGURATION REVIEWS AND AUDITS

The importance of configuration reviews of software was discussed in my first book, *Software Engineering Reviews and Audits*. This step ensures all elements of software configurations are developed and are in control during software integration and test activities. Conducting and performing effective reviews and audits are key before entering into formal software and systems integration.

FURTHER READING

Florac, W.A., and A.D. Carleton, 1999. *Measuring the Software Process*. Addison-Wesley Professional, Boston, MA.

Jameson, K., 1994. *Multi-Platform Code Management*. ISA Corporation, O'Reilly Media, Philadelphia, PA.

MIL-STD-480. 1988 (July). *Configuration Control: Engineering Changes, Deviations, and Waivers*.

Pilone, D., and R. Miles, 2008. *Head First Software Development*. O'Reilly Media, Sebastopol, CA.

Schwaber, K., and M. Beedle. December 2008. Statistical process control for process improvement. *CrossTalk, Journal of Defense Software Engineering*, 50, 833–859.

8

Software and Systems Integration

8.1 INTRODUCTION

The effective methods and processes for software and systems integration require disciplined software design/development practices and test planning, test execution, configuration control, quality management, and reporting of work product testing inside integration facilities to management and the customer. Software technology books, magazines, and articles all over the world are intended to reflect "best practices" from various integration facilities supporting software companies, the military, and aerospace programs and projects. It is the responsibility of management to select effective and responsible test conductors and teams and have in place software and systems integration processes due to the importance and nature of assigned tests to be successful and provide results. Successful software and systems integration objectives are accomplished by:

- Agreeing on and identifying blocking issues
- Assigning responsibility for clearing those blocking issues
- Scheduling dates for responsible teams
- Holding periodic meetings until issues and concerns are closed out
- Evaluating current integration facility schedules

Blocking issues can include open or electronic paperwork; unavailable software test tools; undefined systems under test; and unavailable software and test personnel. The critical roadblock is not having a plan to go forward.

8.2 SOFTWARE AND SYSTEMS INTEGRATION PLAN

The software and systems integration plan (SSIP) defines or references processes and procedures that are used to integrate defined work products, systems or subsystems, and hardware units into a software and systems integration environment. Defined integration processes and procedures include user definitions for software design/development, execution, test, evaluation, and reporting of results during integration activities. The SSIP includes software integration planning and coordination with other formal test activities; risk assessment, product evaluations, configuration management (CM); and other required support activities. An example plan is provided in Appendix B.

A software qualification test occurs where work products are integrated with systems and hardware configuration units associated with other assigned work products. The software qualification and testing verify activities and the responsibilities for assigned programs and project teams. The CMMI® model addresses process integration practices of the CMMI process framework. Process integration practices include:

- Documenting processes for integration activities
- Verifying and validating in integration environments
- Defining requirements for integration environment readiness criteria within the plan
- Guiding to ensure product integration is maintained throughout the project life cycle

8.3 SOFTWARE AND SYSTEMS INTEGRATION FACILITY

The software/system integration facility (S/SIF) is the primary facility for hardware, software integration, and system-level testing. The facility supports software design and hardware equipment integration used to integrate and test integrated software with configured systems. The integration testing in this facility builds incremental delivery of software work products for checkout and use by customers.

8.3.1 Facility Operations

Affected teams build, maintain, and upgrade facility operations for software design/development and tests to be conducted. The environment for software and hardware configurations is established to support early design and test equipment integration. To ensure systems integration facility operations are conducted, systems are integrated, and performance is measured. These activities are conducted for formal verification of system specification requirements. Detailed integration plans are used with plans and test procedures to execute integration testing in the facility.

8.3.2 Facility Configuration

The facility is configured to support design and test operations. Detailed test equipment provides test plans and procedures to be defined for each test conducted. Drawings are documented to lay out the facility configuration and coordinate with hardware, electrical engineering, and hardware quality.

8.4 INTEGRATION SETUP

The integration setup of software and systems work product components occurs within the system, emphasizing interfaces and operations between components, including hardware, software, interfaces, and supporting functions. The work products are integrated and performed incrementally using the process of assembling components, testing, and verification and then assembling more components for setting up the integration activities.

8.5 FORMAL ENGINEERING BUILD

In current states of software design/development and qualification tests, programs and projects become increasingly complex. Formal software engineering builds and releasing of software consumes an ever-increasing amount of time and resources. Software build tools provide a way to automate entire builds, deployment, and quality assurance and release work products to the customer.

8.6 TEST TEAM

The test team is responsible for formal qualifications of a specified system requirement. The test team works inside the facilities' operations with other systems and software personnel. Test teams have always asked software designers/developers about the issues they see in terms that the designers/developers can understand. This can lead to bad feelings between the teams. The test teams and the designers/developers of software need to change their attitude toward each other.

8.6.1 Documentation

The documentation software required for the formal qualification phase defines and documents the progression and interdependency of test artifacts. The documentation required is as follows:

- SSIP
- Integration and installation procedures
- Design documentation
- User and operation guides
- Test and analysis reports
- Compliance documentation or sheets
- Hardware drawings

The requirements verification documentation flow is shown in Figure 8.1.

8.6.2 Roles and Assignments

The responsibility for the conduct of a system test and evaluation is the role assigned to the test conductor and test team organizations. The test team is responsible for preparation of internal processes, test plans, and procedures to ensure verifications meet system and software specification requirements.

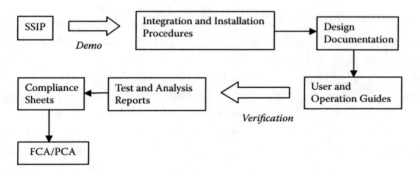

FIGURE 8.1
Verification documentation flow. FCA, functional configuration audit; PCA, physical configuration audit.

8.6.3 Integration Test Processes

The integration test processes are internal activities conducted by test teams to develop test procedures and ensure acceptance testing has been completed at the end of formal qualification testing. A model for integration testing is provided in Figure 8.2.

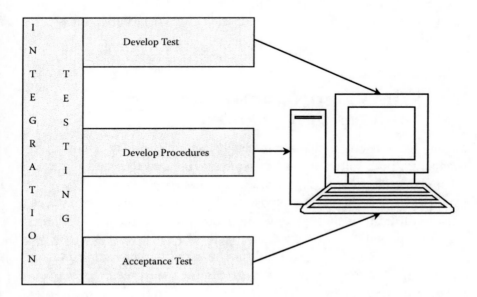

FIGURE 8.2
Model for integration testing.

8.6.4 Problem Discovery

The verification method is used when performing an operation to discover problems and verifying visually; at some point during the test, redlines are applied to procedures. This involves direct observation while using the system in its intended modes and states. In general, demonstrations apply to simple, observable events using pass/fail criteria without test or measurement equipment. Any specification requirement that includes demonstration as a verification method requires verification that equipment/units are working and operations or functional performance is in place. The hard copy or electronically released procedures record the pass/fail status and the marking of a redline when conducting a test and operations.

8.6.5 Problem Reports

The test team approach used to manage and coordinate program and project integration testing problems is documented in problem reports as test problems occur and are discovered. Reviews and monitoring team design and efforts are entered in electronic databases or documents written on hard copy. Metrics are gathered on team efforts to monitor progress, risk, and the resolution of design and test problems during integration testing. Senior, program, and project managers review the metrics to monitor progress for closure of problems reported during integration testing.

8.7 QUALITY PARTICIPATION IN SOFTWARE AND SYSTEMS INTEGRATION

Inside the software and systems integration environment, quality personnel for both software and hardware are required to support integration plans and work products produced by software designers/developers and the test team to ensure software and systems hardware work as one. The test team runs through test installation procedures with the quality team to witness the procedures and verify the media to show that system software works and that results are documented for completion and closed. In military and aerospace programs, the quality team verifies, validates, and approves the media loaded for integration checkout and testing. There is a common approach that the test team will use; redlines applied to an installation procedure are authorized and incorporated in the procedure for the next formal release procedure to support testing.

8.7.1 Quality Checklist

A basic checklist for integration operations is used by the quality team. The checklist required by the quality team will ensure that step-by-step operations are verified and validated and provides a buy-off to work products. The quality checklist will provide:

- Criteria defined from previous audits, plans, procedures, and documented requirements
- Recorded results, including any noncompliances or observations
- An audit report that provides the scope and purpose of the audit, completed checklists, trained personnel, results and lessons learned for future improvements
- Measurement data produced during the audit
- Applicable work products submitted for control in accordance with the software/system plans

8.7.2 Verification and Validation

The verification and validation process addresses work products in integration environments and include selected requirements, including systems hardware and software work product element requirements. It is an incremental process that is performed throughout the software design/development life cycle.

The validation process is performed by the quality team to ensure compliance to plans, procedures, and data inside integration facilities. At times, the software designer/developer, CM, and test teams allow subcontractor participation in a team development environment to receive, capture, and report the assessment of the product's ability to meet the needs of the customer and other teams in the user integration environment.

8.8 LATE NIGHTS, EARLY MORNINGS, AND WEEKENDS

Many late nights and early mornings, the software quality team is required to support software and systems integration activities in the S/SIF. The team is called in at any time to support integration activities. Without the quality team supporting the installations, testing, buy-off, and delivery to other integration lab users or the customer do not occur. Plans for buy-off require quality team verification and validation and approvals applied (i.e., CDs, computer units, redlined test procedures, version documentation, test sheets, etc.).

8.8.1 Software Quality Support

From my experience, if software quality is not available to support software and systems integration operations, the quality team manager receives a phone call or e-mail to ask for support. The quality team at times has no life when it comes to supporting integration activities. Many programs and projects are under pressure when schedules are impacted and depend on the quality team morning and night. My frustration with this is discussed next.

The program and project managers are concerned and worry about delivering a quality product to meet customer expectations. Is this true? Schedule comes first, then quality is somewhere down the totem pole. It is the senior manager's responsibility to guide program and project teams to meet commitments of technical performance, cost, and delivery dates. I know it is tough on senior management to meet all these requirements, but a schedule should be provided that works with all teams that are affected. The quality tasks are everywhere inside the program and projects. These tasks include process/product evaluations, reviews, audits, planning, formal audits, training, and verification and validation of work products to be ready for formal test and delivery. I could continue about the responsibilities required for support. For the senior manager to be responsible for execution of program and project plans, the term *efficiency* is the answer.

When changes are made to planning schedules, include quality teams in the discussions to ensure events in the schedule allow support. I understand that changes to planning schedules change hourly and daily, but also ensure resources are available and expected per updated daily schedules and rescheduled to support the expectations of both senior program and project managers. Late in a day when the quality team is ready to go home, calls are made and require support immediately in the integration facilities to ensure delivery to customers. It is frustrating at times, but programs and projects do expect quality attributes and approvals to be applied.

8.9 BREAK THE MOLD

All military and aerospace programs and projects that are in current operations should learn from the past to improve quality processes and implement sound practices. In other words, lessons learned from past operations inside programs and projects should be discussed and reviews conducted. Teams often state that the current performance is the same process as other programs and projects they have supported. The senior manager along with program and project managers should change the old ways, break the mold, and improve the approach teams should adopt to be more successful. I know that management personnel are not perfect, but they should be able to create a working environment for employees/teams to deliver quality work products to customers consistently and on time.

8.10 THE BOTTOM LINE

At times, it is common knowledge that senior, program, and project managers react to schedule concerns. However, the pressure to have quality teams support these schedule concerns is overwhelming, and program and project managers act foolishly. I apologize for the previous statement, but I have been involved and have witnessed the pressure applied to perform verification and validation with a short timeline and to be ready to release work products to software and systems integration facilities and customers.

The need is to emphasize results, not the time spent on meeting deadlines, tick marks, and schedules. Effective and efficient teams can overcome bad program and project management and schedules. When strategy meetings are scheduled and there is disagreement with the projected schedules, teams or individuals will find other programs or projects to support.

8.11 EFFECTIVE METHODS FOR SOFTWARE AND SYSTEMS INTEGRATION

The purpose of this book is to provide programs and project-effective methods for achieving the success of software and systems integration.

My proposal should be reviewed and implemented in military and aerospace programs and projects. The software industry may be able to review it as an approach to see the capabilities suggested.

The senior managers, program and project managers, and teams that are managing software and systems integration activities are responsible for the integration of work products. The disciplines of system design, software requirements, and design, build, and testing of work products must show continuous improvements in quality throughout the software life cycle.

For development of the work product vision for software and systems integration, key stakeholders must ensure that the definition for work product releases is understood from the start to the finish.

To be effective during integration activities, the following methods for software and systems integration are included:

- Planning
- Communication
- Risk management
- Requirements
- Systems/software design
- Integration
- Execution
- Continuous integration
- Configuration management
- Quality
- Customer satisfaction

8.12 PLANNING

For planning, develop the SSIP and strategy to understand the systems you integrate, including the environment, functions, and constraints. Ensure requirements are testable, operational, and technically realistic. Consider using an integration readiness review plan for operational criteria in the integration environment.

The planning for software and systems integration activities involves everyone from the start, including subcontractors and customers. The programs and projects require integrated processes per released software plans, installation, and checkout procedures. Before conducting software and systems integration, lab operations implement a readiness review to ensure that trained personnel are available and lab environments are ready for integration activities.

8.12.1 Monitor Planning Progress

There have always been ways to track and monitor progress for planning on software programs and projects. The planned resources of building and supporting software and systems integration activities ensures that development time and effort (cost and staff) are in place and receive the go-ahead for implementation.

The projections of not only program and project managers but also higher-level managers are important to ensure and provide broader responsibilities. These managers are responsible for providing resources (i.e., staff, managers, funds, development time, etc.) to enable the pieces to be in place for programs and projects to follow plans and procedures. They will need to know the amount of personnel that will be available months ahead of time to evaluate the ability of the programs and projects to undertake new work.

Key measurement points are called milestones. They occur at points in the software design/development life cycle as suggested in Table 8.1.

Another planning process to consider is to apply statistical control to the software design/development life cycle. Many programs and projects are familiar with this concept to manufacture work products. This tool or concept can be implemented for software and systems integration.

TABLE 8.1

Key Measurement Points

Number	Key Measurement Points	Life Cycle
1	Feasibility review	Higher-level managers
2	Early design review (EDR)	Approval
3	Required design review (RDR)	Specification
4	Code and integration testing	Software design
5	Start of software/systems integration testing	Functional capability
6	Combined operations (software and systems)	Fully functional
7	Full operating capability	Release
8	Monitor level	99% reliability
9	Reliability level	99.1% improvement
10	Continuous level improvement goals	100% "happy customer"

Measurements that fit within these control limits can reflect instability of progress. Some processes are still good, but sometimes the processes fall outside control limits.

Apply the principal of statistical control to a knowledge process, such as having a projection of what is expected to be completed with a low defect rate. The programs and projects at times follow expected plans and procedures to ensure control. If program and project managers are assigning more personnel than planned, there will be an issue of getting back to the proposed plan, or you can preplan the program and project objectives.

The higher-level manager would want to find out what the program and project managers are doing to resolve this issue and if the customer is going to be affected or resources are needed elsewhere, leading to more time to perform integration and cost concerns.

Monitoring project programs will not fall out of the sky but should be managed instantly. The first steps are to:

- Establish baselines
- Collect data from previous programs and projects
- Control work products by collecting basis metrics
- Use teamwork so everyone cooperates to ensure customer satisfaction

8.12.2 Comment

The failure to provide effective planning and coordination in preparation for integration activities will ruin planning and coordination. There is intense pressure when developed schedules require tick marks to be completed and shown to customers. For planned schedules, customers get that warm and fuzzy feeling when milestones are marked off, but to be honest, many software and systems integrations are not complete. Always tell senior, program, and project managers to be up front with teams and the customer to ensure confidence that quality comes first and then schedules will follow.

8.13 COMMUNICATION

Communication encompasses channels for passing information to support interpersonal communications along with feedback and criticism (Figure 8.3). The quality of communication in a program and project is

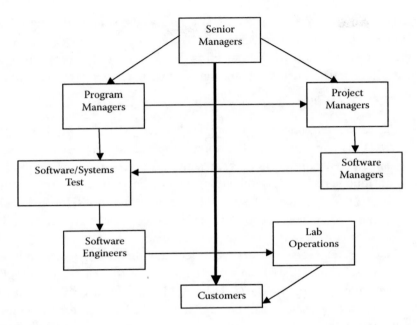

FIGURE 8.3
Communication lines.

directly related to effectiveness. When the time comes for software and system integration activities, it is essential that open information is shared at both technical and interpersonal levels.

The technical level deals with the way information describing the work product or the process is shared. On the interpersonal level, communication deals with feelings about the work product, work relationships, criticism, and personality.

Information in a program and project should be captured and communicated in writing so the understanding and coordination can be shared in the lab environment. In an electronic society in which paperwork is at a minimum, written communication can be considered formal.

8.14 RISK MANAGEMENT

It is highly recommended that risk management is conducted for integration of software and systems so it is continuous and shows the risks that occur during automation in software builds, installations, and test concerns. All risks are documented and reviewed each day during integration activities. The risk management concept is a continuous process of identification and planned team meetings to resolve and answer problems that could be found during integration. The basic process steps are summarized as follows:

- *Risk issues and concerns.* The process begins with the identification of issues and concerns. All integration teams (i.e., design, test, etc.) identify such issues and concerns through peer reviews and discussion of continuous risks so it is known what could have an impact on schedules. When possible, software subcontractor activities are included in team risk reviews. Technical performance metrics are used as the basis for risk identification and assessment.
- *Risk reviews.* Once a risk is identified, the identifying teams review the risk. A risk is rated as belonging to one of three categories: high level, midlevel, and low level. Risks rated as moderate or high require program and project risk management action presented for senior management review. Low-level risks are managed within teams and are reviewed regularly to ensure risk mitigation.
- *Risk management plans.* After a risk is defined and assigned to a team, that team will develop and implement risk management plans and continue to assess risk status until the risk is addressed and closed.
- *Risk monitoring.* For risks assigned to teams, the team provides the risk status using the risk management database. The team lead manages and maintains the database for tracking and reviews. This database generates status charts and reports for programs, projects, and customer reviews.

8.14.1 Risk-Based Integration

Once the program and project managers agree on the estimates to create a plan for risk-based integration, the plan assigns testing based on software design/development and tests. Quality is the level of risk that could affect software and systems integration activities. Risk-based integration is reviewed when analysis is performed to root out software design/development and test defects.

During integration and analysis, the test team allocates development and execution efforts based on risk. The procedures used are based on reactive techniques to detect and sort out high-risk areas. When test results are released, test cases executed, and bugs found during integration, you are able to trace the quality risks.

8.14.2 Risk Integration Standards

Examples of how risk integration standards, including those for quality, apply to embedded software that controls software and systems is identified in ISO/IEC (International Organization for Standardization/ International Electrotechnical Commission) standard 61508. This standard focuses on risks. There are two primary factors that determine the level of risk:

- Likelihood of problems occurring
- Impact of problems that could occur

Technical ideas such as coding and unit tests are the problems that arise when likelihood concepts come into play.

During a program and project, we must reduce risk to a tolerable level when applications are software improvements to a system or hardware unit. We have to build quality from the beginning and not at the end by making defect-preventing actions to software requirements, design/ development, and integration testing. Risk integration standards require software requirements and test design to be structured. Hardware units are visible, but inside these units is software that controls the hardware so it comes alive. The movement of the hardware and software requires multiple levels of testing.

8.15 REQUIREMENTS

The teams define and develop software requirements that are selected for implementation and completion during software and systems integration. Completeness and accuracy for software requirements are verified with key work product developers. The customer should always be included in the definition of the requirements to ensure there is complete and concise understanding for their business needs.

Problems discovered in defining and developing the requirements for software are coordinated with higher-level system personnel and fixed quickly to make sure schedules are not impacted for the release of the work product. Derived requirements come into play when the performances of software are defined and applicable to systems design needs for delivery of the software work product for software and systems integration activities. The definitions of software requirements are documented in the development plan for process and work product standards. The measurement of data and metrics generated are reviewed and verified for completeness by program and project plans.

All software requirements are identified for the automation of builds and installations inside the software and systems integration environment. The software work products are integrated to be correct and reflect continuous improvement.

8.15.1 Evidence of Requirements

Conformance to software requirements shows evidence that program- and project-developed software and commercial off-the-shelf (COTS) or nondevelopment items (NDIs) elements are defined and documented. The documentation of installation procedures shows the evidence utilized for the automation of software build tools. When subcontractors provide software, those elements are identified by approved plans for use during software and systems integration.

8.16 SYSTEMS/SOFTWARE DESIGN

The software design definition is developed and controlled by plans for development or design. The constraints for the software are identified during the start of the software design/development life cycle. There are objectives that are required to meet schedules from the start to the end of software and systems integration.

The software design engineer identifies risks and software restraints that could occur during development and hinder schedules. The software requirements are analyzed for software and systems integration to ensure the software and systems or systems design work together and are viewed to make continual improvements.

8.17 INTEGRATION

Before performing software and systems integration, lab operations implements a readiness review to ensure the lab environment is ready for design and testing. What I mean by a readiness review is to create a high-performance work team (HPWT). The following trained personnel should be in this team:

- Systems designer
- Systems engineer
- Software designer
- Configuration management
- Quality personnel
- Hardware designer
- Subcontractor (if required)

The HPWT will perform a software engineering review and audit of each discipline in the software life cycle to ensure processes are being followed per defined plans, procedures, and expected requirements from the customer. By performing this audit and review, results are reviewed and documented by the HPWT and presented to senior management for discussion with affected program and project managers. It is hoped that before programs and projects are ready to integrate work products into the integration environment, processes are in place and compliant. The program and project managers do not want to hear the following:

- ✓ "I have technical issues and concerns in terms that are understood by other team members."
- ✓ "My software does not work when delivered for software and systems integration."
- ✓ "Processes? We don't have any processes defined and implemented."

8.17.1 Team Coordination

When conducting informal or formal peer reviews, ensure guidelines are understood by the team members from the start to the finish. Team assignment responsibilities define the data collected for each peer review and which tools are used to establish, collect, and store the required data.

Maintain the schedule or plans for new and revised work products to include peer reviews at the completion of the entire work product. The scheduled plan for a peer review should be divided and show incremental peer reviews immediately after completion of each section and should not delay these until the end of the phase when there is limited time for rework. The schedule includes dates for team training.

8.17.2 Plans and Procedures

Before software and systems integration testing is started, test plans, documents, and procedures are required to be released through a documentation release system. All software and systems integration tests are performed based on definitions of elements that are documented and identified during integration.

8.18 EXECUTION

The software and system integration recommendations are to show execution of test-built systems for integration activities and to ensure the builds provided for execution are not broken. Build and test times should be reviewed to minimize problems that occur during the software and system working together for integration activities. Acceptance tests are performed along with the customer as witness depending on the program and project requirements.

8.18.1 Acceptance Test

The acceptance test approach or methods will provide answers to questions asked such as whether the code will do the right thing during software and systems integration. There are many considerations that apply, such as:

- Understanding of the specification
- Efficient integration of design and test
- Improvement of processes
- Decreased regression tests and costs

Requirements define acceptance testing to validate implementation of software and systems integration. The features must have acceptance to involve automation methods, which is my first choice over manual steps. The software and systems or units that are working together are accepted when completion occurs. I feel acceptance testing shows that the code provided by the designer is working as expected and has been peer reviewed and tested. Remember that the software and system are not ready for release to the customer, so perform a readiness review before production.

8.19 CONTINUOUS INTEGRATION

Continuous integration is the automation of build and test processes, starting first with the software code being checked into the computer media library (CML) repository. Teams can assure that the code quality is under configuration control. The automation of the build and testing approach should be implemented to support the following:

- Source code capabilities
- Confidence in builds and testing
- Restoration of previous configurations
- Interaction of compilers and systems design personnel
- Building entire systems from scratch
- Team awareness of builds and test failures

8.19.1 Automation

The automation and generation of software and systems packaging ensure confidence in personnel requests by the integration labs or the customer. Staging builds and tests together supports requirements that could include the package to facilitate the integration process.

Dedicated systems or a hardware platform should run tests continuously. The software designers could hinder and slow testing if this discipline is not applied. There are always new types of tests to be conducted. Test artifacts created by the design team provide a start to build automation.

8.20 CONFIGURATION MANAGEMENT

The definition of CM is a discipline applying an administrative process and direction for work products developed during the software or hardware life cycle. The importance of CM is to identify and document both functional (as-designed) and physical (as-built) work products.

The control of changes to work product plans and procedures during software and systems integration provides and records the information required to manage work products more effectively. Authorization is an important factor; with it comes performing the methods for integration activities in lab environments.

Many programs and projects are required to document a configuration management plan (CMP) to define how to implement sound and effective CM policies. The configuration status accounting (CSA) method for recording and reporting information needed to control work products includes:

- Technical data
- Administrative data
- Design data
- Changes
- As-built products
- As-designed products

By looking at CM, you can see the programs and projects have a better understanding of how each part constitutes the functions such as identification, configuration control, and status accounting and the importance of CM. The CM method is shown in Figure 8.4.

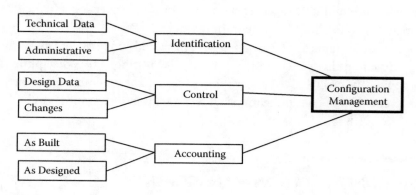

FIGURE 8.4
Method of configuration management.

8.21 QUALITY

The essential practice that teams should follow is quality. The more quality teams adopt, the more successful they will become. At times within the integration activities for both software and systems, teams will see higher-quality software and systems efficiently produced and ready for delivery to customers.

As mentioned in software design, peer reviews and testing can be implemented early in the software and systems integration activities. The concept of automation in building and testing software to ensure systems work together provides the framework for how software design teams write software.

Work toward perfection and show continuous improvement early in testing and evaluations. Teams always think about the best way to improve the processes for testing. Making mistakes or wrong decisions early provides opportunities to learn from mistakes and have them addressed correctly.

Quality attributes ensure teams build and test the right work products. These quality attributes are:

- Build and integrate the correct work product
- Build and integrate the work product to be correct

The relationship between program and projects makes it much easier to understand the requirements early in the software life cycle. At an early point in the life cycle, customers have opportunities to be involved and provide comments and recommendations.

The major goal of effective building and testing for integration is to prevent problems or defects early during the life cycle. Quality is the responsibility of everyone on the team. In software, quality involves communication with other members and discussions of issues and concerns. In testing, teams are required to pursue perfection. To be a success, perform quality disciplines as early in the process and show improvement in all those quality aspects that involve:

- Code quality
- Peer reviews
- Builds
- Testing

TABLE 8.2

Code Quality

Quality	Description	Ensure
Code and unit test	Tested software interfaces	Test coverage on systems
Rules applied	Prevent redundancies	Roles and responsibilities
Team work	Communication	Clarity and understanding
Desktop integration	Code design and test process	Early detection deployment

Code quality and peer reviews are the key elements to ensure software and systems integration are working together. Consistent patterns for using quality solutions solve problems that consistently appear. Experienced software designers incorporate better solutions, and terminology is understood from the start of code development. Make sure conversations are held with team members so there is consistent understanding about code quality. The software architecture allows code to be more manageable and for changes to exhibit good-quality attributes for code development as shown in Table 8.2.

8.21.1 Peer Review Assurance

I am a strong advocate for peer review assurance and will always be. Allowing software engineers to build software often and provide software that is ready for the test team to minimize the number of problems or changes at one time will reduce the risk of defects and errors in the software.

The CMMI process model provides an understanding of integration processes for software and systems integration. The integration process requires a continual emphasis on "repeatable processes." Conducting peer and code reviews defines verification and validation when it is time to audit the software processes performed in the software and systems labs.

Software designers work to requirements, write effective code, and take pride in being considered excellent or exceptional at their profession. The peer and code review process, which is a repeatable process, ensures that other software designers review their code with proper software tools or pair programming compliance to coding standards. Implementation of high coding standards enforces software processes that are implemented with defects or issues are resolved early in software design/development stages, before software and systems integration starts.

8.21.2 Software and Systems Assurance

A fundamental objective of software and systems assurance is continuous improvement in the quality of work products and processes during integration. The assurance and process improvements are achieved by defining, documenting, measuring, analyzing, and improving the development and integration processes to reduce error rates and flow time. The requirements for assurance are documented in plans for software design/development and a software quality plan (SQP). Results are reviewed by integration teams.

8.21.3 Additional Quality Concepts

There are additional quality concepts to review and understand when conducting software and systems integration. If we are unable to explain or say what quality is, then how does anyone know quality exists? Quality does exist. What mechanism or tools helps us understand quality and its meaning and definition? I have heard many engineers and test teams question what quality is and why it is needed.

In software design/development, the quality concepts focus on the degree to which software disciplines are implemented. I have said this repeatedly; if senior, program, and project managers involve all teams in planning to understanding schedules, and contribute and support presented schedules, then the software and test team will have confidence from the start to succeed. The quality concepts will be made easier for the quality teams to support the software and test teams better.

8.21.4 Improving Quality and Productivity

Improving quality and productivity for software basically indicates how well quality meets the requirements and expectations for supporting software and systems integration tasks. This assurance provides adequate, reliable, and efficient software design/development lifecycles. The growth in computer use for software and systems integrations places demands on increasing high-level use and complexity.

The use of effective technology is a means to improve quality and productivity for programs and projects. Military and aerospace companies have looked at software technology as a means to improve quality and better predict costs and schedules required to develop and maintain very complex software systems. Current and future technologies support software design/development processes throughout automation of software engineering practices.

The quality infrastructure is a means to integrate the disciplines that assist systems and software designers, CM, tests, program and project managers, and so on. The communication between team members promotes other team members for improving quality and productivity during software and systems integration.

8.22 CUSTOMER SATISFACTION

Customer satisfaction is the concept of assuring the customer that effective methods for software and systems integration have been compliant and do meet concrete requirement expectations. Many mistakes are made when programs and projects work to poorly defined requirements from the start. Poor execution to system design and software requirements compromises the quality of integration methods to deliver value or causes outright failure. Managers need to ensure customer expectations are understood; unexpected surprises could wreak havoc on program and project schedules.

When poor software and systems integration methods are not effective, program and project schedules lead to major problems with customers.

Everyone wins when there is more focus on the success of the program and project to meet budgets, schedules, and the customer's satisfaction. Strong management, effective team support, and the understanding of what is required to be successful will lead to software and systems that are in alliance with business needs as well as user expectations.

8.23 TAKING THE INITIATIVE FOR CHANGE

How many times have we heard someone say, "That's not my job," and "I don't want to change the way I do things"? It does happen often in the field of software development. We know that things will become better when we resolve issues and concerns, but at times we do not take the initiative to improve or there is not enough time.

The creation of software and systems integration problems has generated change that affects programs and projects. Numerous changes are good, but taking time to get organizations working together can cause problems. When someone states, "We can make integration of work products more effective," we are not perfect. There is an integration environment established to find problems and fix them. People who are working toward improvement shun others, do not listen, and then walk away. The software companies need to fix this in their organizations, even down to the teams supporting each other during integration activities.

When you see that problems occur during activities (i.e., software design/development, testing, CM, quality testing) that are supporting integration, you need to become an expert and take the initiative to change. Improving processes to better fit your work environment will have everyone on the same page. No more walking by and acting like you do not want to help solve problems. That is your job. Change the way you do things now.

FURTHER READING

Augustine, N.R., 1983. *Augustine's Laws*. Penguin Books, New York.

Black, Rex, 2002. *Advanced Software Testing*, Vol. 1. O'Reilly Media, Philadelphia, PA.

Boehm, B.W., 1992. *Software Engineering Economics*. Prentice-Hall, Englewood Cliffs, NJ.

Carnegie Mellon, November 2010. CMMI® for Development, Version 1.3, Improving Processes for Developing Better Products and Services. Carnegie Mellon, Pittsburgh, PA.

DeMarco, T., 1982. *Controlling Software Projects*. Yourdon Press, New York.

Electronic Industries Alliance (EIA), Government Electronics and Information Technology Association Engineering Department, August 1998.

Fagan, M.E., 1976. Design and code inspections to reduce errors in program development. *IBM Systems Journal*, 15(3), 192–211.

Gilib, T., and D. Graham, 1993. *Software Inspection*. Addison-Wesley, Reading, MA.

Grady, R.B., 1994. Successfully applying software metrics. *IEEE Computer*, 27(9), 18–25.

Lipke, W., December 2003. Deciding to act. *CrossTalk, the Journal of Defense Software Engineering*, 21–24.

Putnam, L.H., and W. Meyers, W., 1992. *Measures for Excellence; Reliable Software on time, Within Budget*. Yourdon Press, Prentice-Hall, Englewood Cliffs, NJ.

Syzmanski, F., May/June 2011. Deliver applications that meet business needs. *Better Software*, 34–35.

Weller, E.F., 1994. Using metrics to manage software projects. *Computer*, 27(9), 27–33.

Whitaker, K., 1994. *Managing Software Maniacs*. Wiley, New York.

Yourdon, E., November 1994. Software metrics. *Application Development Strategies*, ISO/IES Standard 61508.

9

Software Subcontractor

9.1 INTRODUCTION

This chapter describes the methods that are performed by a software sub-contractor to provide the necessary support and employment that would benefit military and aerospace programs and projects. The software sub-contractor can be hired for program and project planning, configuration management, quality issues, software design/development, testing, and execution of activities or tasks related to the delivery of software work products to customers. The activities performed are in accordance with a purchase contract, and the software work products are delivered to satisfy and comply with specified acceptance and delivery requirements.

9.2 PROGRAM AND PROJECT SELECTION

The selection of a subcontractor for software programs and projects is important due to the expectations and technical work disciplines required. Selecting a subcontractor for contracted software design/development is the responsibility of program and project managers.

The program and project managers provide the direction to perform various job tasks related to the day-to-day software design/development to be delivered for subcontracted work products. There is competition from other subcontractors to obtain assigned work.

The planning for the subcontracted work is performed during a program and project start-up once decisions are made to employ outside support. The outputs from this action are the responsibility of a software subcontractor plan (SSP), but specific tasks can be assigned to this plan or to other teams depending on organizational requirements or needs. A statement of work (SOW) will list subcontractor requirements.

A list of requirements, expectations, and interfaces between the program and project are documented in an SOW. The goals of selected software processes are given to the subcontractor per the direction of program and project managers in order to flow down plans for required tasks during a software life cycle. This permits the software subcontractor to abide by internal software processes that require objective evidence to be reviewed and ensure goals are being accomplished.

The selection and search for a subcontractor is required by program or project plans or the subcontractor program or project managers. Support may include teams or the coordinated experience of software engineers for understanding all aspects of a technical approach, evaluation, cost, estimates, and software-related tasks as needed.

For many years, I have been in the role of a customer and subcontractor. When you are the customer, allow the subcontractor to present his or her case and the reason it was selected and under contract. Many WebEx and face-to-face meetings are needed. The subcontractor will feel comfortable and know the purpose for selection. Many subcontractors are uncomfortable when presenting what their company can provide to customers. Give them a chance and let them relax. Be positive when presentations are provided for review and ask questions to see what the answers will be. Be positive; some companies will not select you as a subcontractor but will move on to the next customer. The learning process will benefit upcoming subcontractors for understanding what customers will want to hear. Again, *be positive.*

9.3 SUBCONTRACTOR APPROACH

The software subcontractor is an integral part of the team for software work product development. Ensure progress by the appropriate subcontractor for software activities and progress is based on agreed evidence of completion. The program and project organizations, with the support of senior managers, oversee the subcontractor's work quality, engineering, and systems and are in continuous support as required by contract.

9.4 SOFTWARE SUBCONTRACTOR PLAN

The SSP provides the direction for the subcontractor hired and the program and projects for understanding the requirements and guidelines for both organizations. Each software subcontractor is responsible for configuration management of his or her software at the subcontractor's own facilities in accordance with the plans and procedures and abides to the standards, processes, and procedures of the program and projects under the signed contract.

The plan identifies the program and project managers' approach to manage the required subcontractor's effort. This plan will describe:

- The task for each subcontractor
- The processes for configuration management and quality audits
- Requirements
- Risk management
- Configuration management
- Schedules for delivery of work products

The plan includes support responsibilities and explanations for subcontract development and how the subcontractor will be managed. The associate subcontractors and major subcontractors are defined in the plan along primary roles that specify how the program and projects interface and measure performance.

All subcontractor deliveries to a customer require receiving and inspection of the software work product. Software could be delivered as media or electronically. The data are stored with configuration management for accountability and checkout for use. Hardware deliveries are received as packages or boxes. This process is also defined in the plans or internal procedures.

9.4.1 Software Audits

The software audit is comprised of program and project reviews to be conducted at subcontractors' site of business by defined dates as documented in the SOW. The subcontractor software plans and procedures are audited per defined and documented audit methods to trace information for software requirements to/from applicable test cases/test procedures and per the signed contract.

The subcontractor plans that are audited must ensure the software test environment performs its intended function and meets contract requirements. The purpose of this task is to ensure that the software under test is qualified on acceptable test tools.

Established subcontractor process audit criteria are prepared and provided to the subcontractor before the audits are performed. An audit checklist is provided, and audit questions will be filled out and then presented to the customer. The agenda and participants are identified using a defined audit process applicable per the contract. Involvement in the software first-article inspection (FAI) is the approach; I highly recommend programs and projects prepare an FAI checklist and deliver it to the subcontractor to provide answers to questions before performing the audit. The results save time and cost along with traveling all over the world. We have the technology to perform WebEx or telecommunications capabilities to discuss audits and action items instead of being at the subcontractor's facility.

9.4.2 Audit Checklist

Appendix C provides an example of an audit checklist I developed and can be used with subcontractors as well.

10

Software and System Delivery

10.1 INTRODUCTION

It is important to make the right decisions before delivery of software and system end items or hardware to the customer. At times, schedules become the priority before quality, and the lack of confidence in the customer will have an impact on future working agreements and contracts. Make sure that systems design, program and project planning, software requirements, software design, and software and systems integration are successful and that every step or milestone has quality built in during the software design/development life cycle. Knowing problems still exist, senior, program, and project managers do not show a tick mark to show schedule accomplishments to customers. Be honest and up front with the customer and know that quality comes first; then, schedules provide the road map for the teams to produce effective work products.

The effective methods for software and systems integration will provide assurance that customer requirements are met before any thoughts about hurrying delivery. Stay the course and do not deviate from the plan. Before delivery of software and systems to customers, the following are important:

- Software media and data verification and validation are complete.
- Software documentation is released and ready for delivery.
- Necessary FAIs (first-article inspections), FCAs (functional configuration audits), and PCAs (physical configuration audits) are conducted, and all action items are closed.

10.2 SOFTWARE MEDIA AND DATA DELIVERY

Software media are identified and labeled per an identification scheme. The identification and media labeling should be in accordance with security requirements for a program and project as presented in a defined and documented security plan.

Marking information can be displayed electronically for all software media and on the exterior of the physical media (i.e., disk sets, DVDs, CDs, etc.) containing software. Software work products are identified in program and project development plans. An identification approach is assigned to all released software and the accompanying software documentation. An example definition of a software part number can be used as follows:

- *Master*: Stays in the CML (remember this all designers/testers)
- *Copy*: User checkout for software design, test, troubleshooting, and the like
- *Disaster copy*: Keep off-site for retrieval due to lost or destroyed media

10.2.1 Software Documentation

Software documentation provides defined and documented releases for various levels of software and systems integration. Software documentation can be used as follows:

- Systems engineering plan (SEP)
- Development plan (DP)
- Software configuration management plan (SCMP)
- Software test plans and procedures
- Software and systems integration plan (SSIP)
- Quality plan (QP)
- Documentation for version control
- Build and installation procedure

10.2.2 Version Control Documentation

There should always be documentation to provide version control that will identify and describe software versions of existing work products. This type of documentation, such as a version control document (VCD), is used to release, track, and control the software versions at the software and system levels.

10.2.3 Build and Installation Procedure

The build and installation procedure describes in detail how to build and install software for systems integration. The configuration management team, with input from software designers, develops build and installation procedures for software and systems integration builds. The CM organization inside a program and project is responsible for the development, control, and release of build and installation procedures.

10.2.4 Delivery Package

The software and systems delivery package consists of software media and documentation associated with the version of the software, printed copies, and identified system work products or hardware packages. Contractual software and systems delivery requirements or agreements ensure delivery to customers.

Software deliveries are used to meet contract delivery requirements or obligations the program has agreed to accomplish. Senior management and program and project managers along with teams provide coordinating and delivering the package document program and project schedules as a completed milestone.

10.2.5 Final Software and Systems Delivery

The final software and systems delivery is the last delivery once program and projects have completed the FAI and FCA/PCA. The following steps provide an example:

- Integration testing is completed; results are acceptable and meet technical requirements.
- Customer accepts the as-designed and as-built product and associated documentation.
- A customer notifies that software or system product is received.

In all phases, the delivery system provides processes and procedures to get things done right. By strengthening these delivery systems, programs and projects can sustain continuous improvement. If these systems are

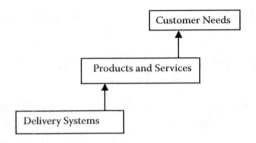

FIGURE 10.1
Customer needs.

ignored, you run a risk of implementing ineffective delivery systems. With a good system in place and constantly improved, the chance for improving work products and services increases.

All programs and projects have customers. Always make the customer happy. The customer can be part of the software and systems integration activities. This environment serves the needs of customers as shown in Figure 10.1.

To survive the global market, programs and projects must continuously improve their work products, services, and delivery systems. Configuration management streamlines the ability to identify and refine the requirements during software and systems integration and through the entire software life cycle.

Business goals are accomplished when delivery systems are created to support those goals. The delivery system must be effective and efficient. The right way to build these systems is to comply with the business process infrastructure.

Programs and projects create work products and services to meet a customer's needs and need to develop the right delivery systems. Always improve the delivery system to improve the future customers coming your way.

10.3 FIRST ARTICLE INSPECTION

An FAI for software is conducted to examine subcontractor production units and if the software is ready for delivery to the customer. If the subcontractor cannot complete all of the tests for the production unit, the FAI will not serve its purpose. Military or aerospace companies are doing a subcontractor a favor by allowing the subcontractor to use his own environment for the formal test. Why should a company receive a production unit without the applicable documentation supporting a formal acceptance test? If a subcontractor cannot find any way to complete a test, call the software FAI off until the subcontractor is ready for the FAI to be conducted.

Military and aerospace programs provide detailed test cases and regression analysis for fixing problems, including the following:

- The test stand description that confirms the components (part numbers) matches the conform test stand.
- The test cases/reports are rerun for the software changes.
- The analysis shows the fix will not affect other parts of the system.
- The new software has been tested through the test stand and witnessed by a company engineer to confirm the new implementation is working properly.

Subcontractors work with the customer's engineering teams to finalize the regression analysis. Software FAI checklists include:

- *Verification requirements*: For embedded software, state the approved production test procedure (TP), including software version. State that testing is completed, results are acceptable, and the software meets technical requirements.
- *Data package*: The product released and approved and software meet requirements allocated to the software. All requirement deviations are recorded and approved. Software life-cycle data comply with plans and standards and are controlled with software plans. Software and life-cycle data are in a controlled software library and archived both on-site and off-site.
- *Version control document*: Traceability is established to system name, subcontractor system part, and document number. Source code components for the software are identified, and problem reports are resolved since the last product baseline was identified. Version documentation identifies the software life-cycle environment and operating software. Software and verification tools are identified.
- *Verification process*: Verification testing is conducted, defined, and controlled. Verification and validations are complete, and discrepancies are captured in problem reports.
- *Product release*: The executable object code was generated from released, controlled, archived source code and released procedures.

The released software is identical to the tested software. If not identical, the differences must be specified and justified. For loadable software, the released object code loaded on media is identified in compliance with loadable software standards. The displayed software configuration identifiers or checksums match high-level and version-controlled procedures. For loadable software media, the label indicates acceptance by quality/configuration management teams.

- *Acceptance test*: The acceptance test environment is defined and controlled, and the TP is approved and released and under configuration control.
- *FAI completion*: There is evidence of software acceptance, and action dates for action items are defined; the software FAI checklist is provided to stakeholders.

10.4 FUNCTIONAL CONFIGURATION AUDIT

The FCA verifies that the work product performance complies with the hardware, software, and interface requirements specification (IRS). It is required that the test data are reviewed and verified, showing that the hardware and software perform as required by the functional and allocated configuration identification. The FCA provides the prerequisite to acceptance of a configuration item. A technical understanding is a requirement to be accomplished concerning the validation and verification per the TP concerning software. FCA activities involve the following:

- Verification that the work product performs to required configurations.
- Major or minor engineering changes are released.
- A product and configuration baseline is established.

10.5 PHYSICAL CONFIGURATION AUDIT

The PCA identifies the baseline for production and acceptance of the work product, both hardware and software in Table 10.1. The PCA verifies that the as-built configuration correlates with the as-designed product

TABLE 10.1

Configuration Baseline

Baseline	Description	Documentation
Functional baseline	System design requirements	System design: high- and lower-level documents
Allocated baseline	System and software requirements	Systems design documents Requirements documents Design documentation System design: lower-level documents Software test documents
Product baseline	Aggregation of internal systems components into software work products	Operations and maintenance documents System and software design documents Version control documents (VCDs) Software user documents, manuals, and procedures Systems and software installation procedures
High-level product baseline	Aggregation of systems design and software high-level documentation into a component	Software user documentation, manuals, and procedures Systems and software installation procedures
Lower-level product baseline	Aggregations of systems design and software and lower-level documentation into a component	Software user documentation, manuals, and procedures Systems and software installation procedures

configuration, and the acceptance test requirements are comprehensive and meet the necessary requirements for acceptance of the production unit. Equally important, it demonstrates that management systems for quality, engineering, and configuration management information accurately control the configuration of subsequent production units. Incremental and progressive audits are performed on systems and major assemblies to build up to the PCA.

PCAs have an option to be conducted concurrently with the FCA. Extracts from the previous FCA audit plan are made available to the team. Quality assurance and senior management ensure available budget, and engineering personnel execute per the PCA audit plan. Metrics captured for the FCA are similar for the PCA for compliance and review during the audit. The PCA entry checklist is provided in Table 10.2.

After reviewing the materials presented, including known issues and subcontractor status, the recommendations by the PCA team are

TABLE 10.2

PCA Entry Checklist

Entry Checklist	Yes	No	Achieved
Kick-off meeting is held to define roles and responsibilities for conduct and performance of formal audit.	X		Roles and responsibilities defined and used as guideline to support the formal audit.
Delivery is received of data packages (i.e., plans, procedures, drawings, system designs, media, logs, etc.) to support the formal audit.	X		All data packages and artifacts are provided as requested by the formal audit team.
Approved nomenclature and terms are agreed on as applicable during formal audit.	X		All nomenclatures and terms are in accordance with the formal audit and understood by the formal audit team.
List of current deviations, waivers, and higher-level changes are requested or approved.		X	Action item 1
Approved requirements documentation identifying the baseline is available.	X		Approved requirements documentation identifying the baseline is provided to the formal audit team.
As-built records are complete and released.		X	Action item 2

AI	Description	ECD
1	List of current deviations, waivers, and higher-level changes requested or approved are not ready for use by the formal audit team.	mm/dd/yyyy
2	As-built records are not completed and released for use by the formal audit team.	mm/dd/yyyy

Note: AI, action item.

favorable, so the PCA may proceed. PCA execution and the metrics will be completed, and the schedule for the PCA final meeting is coordinated with the customer.

PCA activities are as follows:

- The as-built configuration correlates with the as-designed product configuration.
- The acceptance test requirements are determined per quality assurance.
- Engineering changes are released.
- The final product baseline is established.

Appendix D provides a checklist template that can be used to conduct and perform software PCAs at a level of understanding for required contractual documentation, media, and facilities setup for integration. The military and aerospace programs and projects utilize the concepts and checklist steps to ensure delivery to a customer and hand off the work product. At the closure of the PCA, the statement to the customer is: "It is all yours now"—have fun.

FURTHER READING

Electronic Industries Alliance (EIA), Government Electronics and Information Technology Association Engineering Department, August 1998.

Humphrey, W. 2006. Sweet Predictability. *World of Software Development*, 14(2), 14–17.

Keyes, J., 2004. *Software Configuration Management*. CRC Press, Boca Raton, FL.

MIL-STD-480. 1998. *Configuration Control: Engineering Changes, Deviations, and Waivers.*

MIL-STD-973. 1992. *Configuration Management April 1992.* This military standard is approved for use by all departments and agencies of the Department of Defense (DOD).

MIL-STD-1521B. 1985. *Technical Reviews and Audits for System, Equipment, and Computer Software,*

11

Product Evaluation

11.1 INTRODUCTION

The product evaluation is an integral part of program- and project-level activities that is scheduled and performed by quality software personnel on an ongoing basis. These evaluations form the basis for certification that software design/development activities have been performed in accordance with program and project plans and procedures and are in line with required quality requirements.

11.2 QUALITY ASSURANCE

The quality assurance team or organization for software provides product evaluation processes and specific quality assurance for effective software engineering methods and use of software tools. The quality team does ensure compliance to software design/development standards and control of work products and changes. The practice of quality management is applied throughout the software design/development processes. It is important that quality attributes become the responsibility of everyone supporting development environments in software companies and military and aerospace programs and projects. The management of quality for software activities is summarized in engineering reviews, change control, or subcontractor audits and compliance to standards, verification, and validation.

11.2.1 Software Quality Plan

The software quality plan (SQP) describes and documents the software quality assurance roles and responsibilities to ensure that programs and projects are following procedures and processes defined per development plans and other applicable standards.

This plan provides a documented process for assessing software life-cycle processes and their outputs to obtain assurance that objectives are satisfied; deficiencies detected and evaluated, tracked, and resolved; and software work products and software life-cycle data conformed to intended requirements.

Quality reviews/internal audits are performed to ensure compliance with released processes and AS9100C for measurement, analysis, and improvement activities to be conducted; senior managers must adhere to this review and audit. These activities include examining:

- Program and project artifacts
- Released processes and procedures
- That organizations meet the configuration management requirements of Electronic Industries Alliance (EIA) standards
- That organizations meet the requirements of AS9100C

11.2.2 Software Engineering Process Group

A quality assurance internal organization for software has a relationship with the software engineering process group (SEPG). In programs and projects, this is a group or team with assigned responsibilities for monitoring software process activities.

This group operates with appropriate functional areas and software personnel and serves as the software process improvement center for systems and software engineering. The program and project managers, focal points, configuration management, quality team, and team members make up the group and perform the following functions:

- Evaluate company and program best practices to promote these best practices to document software engineering processes and procedures
- Establish and use a process for receiving, evaluating, and acting on and reviewing results for proposed processes, procedures, and technology changes

- Develop software processes and procedures for the entire life cycle that comply with software engineering standards, comply with contractual requirements, support ISO (International Organization for Standardization) 9001 and AS9100 requirements:
  - AS9100C quality management system (QMS) software requirements
  - AS9100D QMS audit requirements
  - Obtain software program and project manager approvals of software engineering processes and procedure changes released for use by engineering

11.3 PRODUCT EVALUATION SCHEDULE

The quality organization or team performs product evaluations to ensure software design/development. Test and integration phases are conducted per a product evaluation schedule (Figure 11.1).

11.3.1 Senior Managers

Senior managers lead lower-level program and project managers. All teams are affected when the pressure is on to produce, show results with no mistakes, and deliver work products on time. The customer is always right because if programs and projects do not deliver work products on

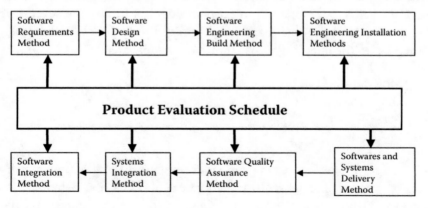

FIGURE 11.1
Product evaluation schedule.

time, panic and chaos occur. That is when you see the importance of quality factors come into play, and milestones are achieved and are a success.

11.3.2 Program and Project Managers

The program and project managers are required to provide a vision statement and describe the benefits, goals, and objectives for developing a work product or system for delivery. The work product or system should be the highest priority for customer requirements to identify the importance of business criteria and requirements. When communication or coordination is not provided to the responsible team, the failure point occurs. A defined vision statement is an example list that program and project managers are responsible for producing; it includes issues related to the following:

- Team objectives
- Risk mitigation
- Issues and concerns
- Root cause (RC) analysis
- Corrective action (CA) plans
- Significant accomplishments

The focus of successful program and project managers is the team, processes, and work product. A manager who fails to communicate early in the software development process could pay a heavy price by making wrong decisions. The manager who pays little attention does run into risks if competent methods and tools are not made available to the team. Having a solid program and project plan does not hinder the success of the program and project. Team members need to be highly skilled software people. The talents that the team should apply are:

- Motivation
- Organization skills
- Attention to business goals
- Work ethics

The objectives for product evaluations are established, and solutions, technical constraints, and alternative solutions are always a consideration. The system and software/design developers, along with test, configuration management, and quality teams, define objectives. The product objectives identify the goals for design and development of data, which provides functionality in a quantitative manner. The program and project managers select the best approach with consideration of constraints imposed by delivery deadlines, budget issues, available team members, and technical solutions to problems.

Processes that are understood provide the framework to implement effective plans and procedures for software design/development activities. The framework details the number of tasks, milestones, and applied quality factors to enable activities to adapt to requirements for programs and projects. Aspects of configuration management and quality assurance are important to the independent nature that occurs during the process.

11.3.3 System and Software Team Participation

Independence of quality support is ensured by the separation of reporting chains to a level that is independent of a program and project. This independence for the quality team increases the objectivity of the product evaluations, which allows the team to provide a better oversight function and involves the system and software teams. The quality team participation in formal reviews does provide team support for product evaluations, and reviews are scheduled and coordinated within team and life-cycle activities.

Each evaluation or review conducted by the quality team generates reports containing the status of the audit, any findings, observations, and recommendations. Compliance, noncompliance, and opportunities for improvements are documented in quality reports and tracked for closure with support from program and project managers.

11.4 ARTIFACTS

Development of system and software work products yields artifacts, including specifications, plans, and procedures. Artifact information associated with quality product evaluations includes software configuration records, testing records, and other artifacts associated with activities, including:

- Audit records (i.e., electronic or paper) associated with product evaluations
- Audit and product evaluation checklists
- Audit results and audit reports

11.5 AUDIT FINDINGS

Quality organizations or teams utilize criteria audit finding derived from software plans and internal procedures to perform scheduled product evaluations. Product evaluations include:

- Review of plans and procedures that oversee programs and projects to determine and select appropriate product evaluation criteria
- Review and analysis of the results of previous product evaluations
- An assessment of whether implemented processes are compliant or noncompliant
- Identification of issues or an opportunity for improvement
- Additional product evaluations required

The results of each product evaluation are recorded in evaluation plans and are added into databases recording summary information from performing an evaluation. The quality team uses these evaluations to indicate if processes are compliant, noncompliant, or there is an opportunity for improvement. Other information may be developed as needed to address team activities or processes. Information on CAs for follow-up of deficiencies reviewed and discovered during product evaluations are also maintained in the applicable databases.

Quality team members generate reports weekly or monthly and are provided to managers.

The weekly/monthly team report data are a record of any issues, noncompliance, opportunities for improvement, and so on that were identified and documented during product evaluations, and the status of all open items requiring a CA is recorded and logged into required databases. Metrics are collected weekly or monthly.

11.6 CORRECTIVE ACTIONS

A corrective action is required to eliminate or mitigate the cause of a detected nonconformity or other undesirable situations to prevent recurrences during product evaluation. Two types of a CA initiated are as follows:

- The root cause (RC) requires RC analysis and actions taken to address the analysis.
- The immediate action (IA) is taken to address a direct cause and prevent recurrence of a specific nonconformity.

Collective analysis is performed periodically to monitor adverse trends of detected nonconformities and undesirable situations that may not have been addressed by an RC or IA analysis. Results are recorded and provided for senior manager reviews of progress/status and overall process performance. The review may be accomplished through a corrective action board (CAB).

Appropriate reviews of the CA and IA should be conducted regularly to:

- Monitor progress/status and overall process performance (e.g., timeliness, efficiency, effectiveness)
- Review adverse indicators and trends
- Resolve issues or elevate them to the proper forum or level of management for resolution

Quality gates come into play to ensure process and work products are compliant (Figure 11.2).

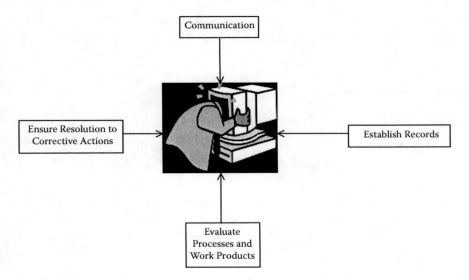

FIGURE 11.2
Quality gates.

11.6.1 Corrective Audit Plan

The product evaluations performed on deliverable work products produced during each product development phase ensure compliance to requirements. Problem definition, investigation, RC analysis, software design/development, and implementation are verified in the corrective audit plan (CAP).

11.7 QUALITY METRICS

In multiple programs and projects, software engineering is required to produce high-quality systems within a defined time frame to ensure reaching customer expectations. To achieve this requirement, effective methods and current software tools should be used to produce quality metrics. Trained senior managers and program and project managers measure if high quality is to be reviewed. The quality of a system, application, or work product is only as good as the requirements that describe problems and test results that are discovered early in the process.

The quality metrics collected by software engineering to ensure program and project's delivery schedules, what is in-work, and completed product evaluations. These types of metrics provide an indication of the effectiveness of a software engineer, test teams, configuration management, and software quality personnel. An example of quality metrics is shown in Figure 11.3.

11.8 QUALITY MANAGEMENT SYSTEM

The QMS is required to have processes documented and executed with knowledgeable people and teams. At times, metrics are reviewed and monitored to ensure processes are showing improvement.

Customer focus is QMS and provides the framework to say; what you do, do what you say, prove it, and show improvement. The standards for QMS are AS9100, AS9100C, AS9100D, SAE AS9110, and ISO 900, which are the models for:

- Quality requirements
- Design and development

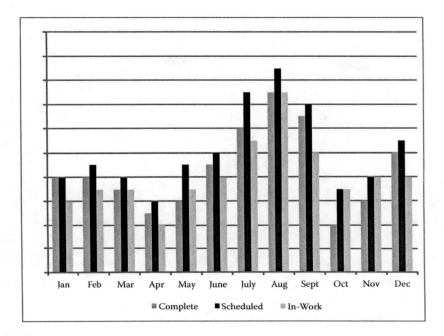

FIGURE 11.3
Quality metrics.

- Production
- Software and systems installations

The support of CMMI® provides the basis for conducting product evaluations, reviews, and audits to ensure compliance to requirements as shown in Figure 11.4.

Measuring quality does ensure a program's and project's operational goals are successful during the software design/development life cycle. It is so important to measure software engineering processes and determine whether programs and projects are consistently improving. If quality metrics are not used, then there is no way to determine if any improvement is within sight. If there are no improvements, it means you are lost, confused, and out there somewhere during software design/development activities.

By evaluating productivity and quality, teams and management establish goals for improvement of the software processes. Using quality metrics, baselines become more manageable and benefit the program's and project's processes to make sure work products operate at a higher level of consistency.

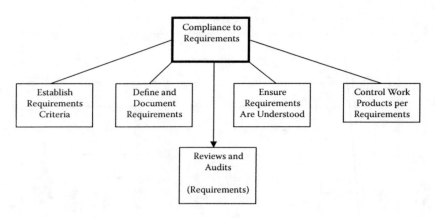

FIGURE 11.4
Compliance to requirements.

Numerous programs and projects run on collected data. The goal is to make software work products and processes better and more cost effective. Data are the key to important decisions. Many problems are linked to using poor-quality data, including:

- Poor estimations of program or project costs
- Not meeting schedules
- Not having effective staffing of personnel
- Flawed software architecture designs
- Poor design decisions
- Ineffective testing decisions

The information on quality measurements is everywhere, and programs and projects have different definitions and requirements when it comes to the quality of data measurements. Mistakes contribute to ineffective information for quality improvement. The most common errors in measuring data are the following:

- Not understanding the measurement goals
- Teams not involved in quality improvement decisions
- Management sponsorship for measurement of quality metrics or data
- The collection of poor data inside program and project development
- Poor data collection and analysis

To help the need for quality and metrics, the Software Engineering Institute (SEI) CMMI version 1.3 for the development of measurement and the analysis method can be used. This method allows programs and projects to perform quality product evaluations at a high level and to establish metrics in standards and best practices.

11.9 SOFTWARE PROCESS

The software process is effective and followed if programs and projects have the discipline to enforce process needs and to follow these processes at all times when product evaluations are conducted. Every software program or project encounters problems as it moves through the life cycle for software design/development.

Proven solutions to these problems should be addressed early and fixed quickly. Established process patterns show a consistent method for explaining problems in the context of software processes.

11.9.1 Software Process Assessment

The existence of the software process assessment in programs and projects does not guarantee that software work products will be delivered to the customers on time and will meet their needs. The process itself can be addressed and assessed to ensure that the process meets a set of basic criteria to show successful software engineering practices will lead to effective software and systems integration to improve processes. The approach for software process assessments and audits is proposed as follows:

- Standard CMMI Appraisal Method for Process Improvement (SCAMPI)
- ISO 9001
- QMS
- AS9100C

11.9.2 Software Reviews

Software reviews provide the framework and detailed requirements for verifying/validating design/development efforts. It is important that performing reviews that are successful will ensure achievement in all specified requirements for software design, test, configuration control, and

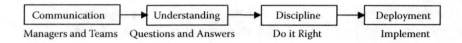

FIGURE 11.5
Process improvements.

quality to released configuration baselines. Reviews improve the individual and team efforts in maintaining a professional setting where software is developed for profit, cost reduction, and service quality improvement.

There is no clear-cut approach to performing software reviews for multiple companies and military and aerospace programs. At times, ideas, suggestions, standards, and concepts are adopted or implemented to improve the quality of software management, software design/development efforts, subcontractor deliveries, and customer expectations. It is frustrating and confusing when it is time to perform software internal and formal reviews. We always ask what is required, who needs to participate, and what results do we receive for performing software reviews. The answer is quality for delivery to satisfied customers.

11.9.3 Software Process Improvement

The process assessment concept proposed in Section 11.9.2 brings order to the stress and chaos for design/development activities, which can lead to failure of software and systems integration. There are no easy answers, but there are alternative options available to system and design/software engineers, test teams, and configuration and quality organizations. Software process improvements become successful if the model for process improvements in Figure 11.5 applies.

11.10 STRESS MANAGEMENT TECHNIQUES

Just in case organizations or team members have rough days and are struggling with processes and other team members, there is a seven-step management technique I would recommend, which does work:

- Picture yourself near the ocean.
- The ocean is blue and crystal clear.
- Birds are flying by and chirping.

- You are the only one there and in total seclusion.
- There are soothing sounds, and the air is filled with serenity.
- You can easily make out the faces of the team members under water.
- See—you are smiling.

11.11 SOLVING QUALITY ISSUES

At times, there are problems with the delivery of work products to software and systems integration facilities and to customers. The quality engineers solve process issues and concerns; they are unable to fix code and test software but participate in peer reviews and witness integration test activities.

The quality engineers provide assistance and help program and project managers look good and become successful, but they need to listen and understand the roles and responsibilities of quality engineering. The quality factors are essential and important to understand.

FURTHER READING

AS9100C. 2010. *Quality Management Systems—Requirements for Aviation, Space and Defense Organizations.*

AS9101D. 2010. *Quality Management Systems, Audit Requirements for Aviation, Space, and Defense Organizations.*

Carnegie Mellon, November 2010. CMMI® for Development, Version 1.3, Improving Processes for Developing Better Products and Services. Carnegie Mellon, Pittsburgh, PA.

Kant, R.K., 2006. *Software Engineering Quality Practices.* Taylor & Francis, Boca Raton, FL.

Pressmen, R.S., 2005. *Software Engineering, a Practitioner's Approach*, 7th edition. McGraw-Hill, New York.

Wellins, R.S., D. Schaff, and K.H. Shomo, 1994. *Succeeding with Teams*, Lakewood Books, Minneapolis, MN.

Appendix A: Acronyms and Glossary

Acceptance Criteria: The criteria that a system or component must satisfy to be accepted by a user, customer, or other authorized entity.

Audit: An independent examination of a work product for software or set of work products to assess compliance with specifications, standards, contractual agreements, or other criteria.

Baseline: A specification or product that has been formally reviewed and agreed on and can only be changed through formal control processes.

Build: Operational version of a software product incorporating a specified subset of capabilities that informal and formal work products include in multiple configurations.

Build Engineer: A role for an integrator to provide a build strategy and build tools to create software baselines ready for test and integration.

Build Request: Requests to software configuration management to provide software builds for software systems and use for computer labs and support the formal test.

Capability Maturity Model Integration: Collection of process models and methods for use in new disciplines to be integrated for organizational structures.

Certification: A written guarantee that a system or computer program complies with its specified requirements.

Change Control: The processes by which a change is proposed, evaluated, approved or rejected, scheduled, and tracked.

Code and Unit Testing: Written routines that specify the disciplines where data is represented, design is understood, data is understood, software developer notes, test results.

Computer Data: Data available for communication between or within computer equipment.

Computer Language: A defined structure devised to simplify communications with a computer.

Computer Program: A sequence of instructions for processing authorized changes per a computer system.

Computer Program Library: Provides permanent archival storage for software and related documentation.

Computer Software Component: A logical or functional grouping of software units to which the configuration management tools assign a unique name, supplied by the software designer.

Computer Software Configuration Item: An aggregation of software designated for configuration management and control.

Computer Software Units: File names that consist of work product followed by a descriptive software unit name (abbreviated to meet operating system and character limits) plus version numbers.

Configuration Audit Plan: Plan that is used as configuration audit steps and instruction for internal and formal audits performed.

Configuration Item: Individual or significant part of a system that ensures changes are controlled, configuration status accounting records maintained, and audits performed to verify product configuration.

Configuration Management: The process of identifying and defining the configuration items in a system, controlling the changes and release of these items throughout the system life cycle, and recording and reporting the status of change requests to verify completeness.

Configuration Management Plan: It establishes a well-documented and established configuration management and control practice for the overall software system baseline to be maintained in the software design/development and system integration facilities and to reflect a set of controlled and configured products and documentation.

Configuration Status Accounting: The recording and reporting of information needed to manage a configuration, including a list of approved changes and documentation.

Control Files: Files stored in a computer and controlled and password protected.

Corrective Action: Repair detected nonconformities in defined processes or other undesirable situations to prevent recurrences.

Corrective Action Board: A review board to detect nonconformities and undesirable situations that may not have been addressed. Results are recorded and provided for senior management reviews of progress/status and overall process performance.

Corrective Audit Plan: Problem definition, investigation, root cause analysis, software design/development, and implementation are verified.

Critical Design Review: A formal meeting at which a critical design is presented to the user or customer for comment and approval.

Data: A representation of facts, concepts, or instructions suitable for communication and interpretation for processing.

Data Management: Controls the acquisition, analysis, storage, retrieval, and distribution of data.

Defect: The aspect of software design/development coding issues when a work product diverges from product designs.

Delivery: The point in the software design/development life cycle at which a product is released to its user for operational use.

Design: The purpose of defining the software architecture, components, modules, interfaces, and data for a software system to satisfy specified requirements.

Design Phase: The period of time in the software life cycle for software design/development.

Development Folder: The detailed design for newly developed software capabilities will be documented in the development folder and reviewed during internal detailed design reviews.

Development Plan: Establishes the plan for development of software during the phase of the program. This plan establishes system-level engineering life-cycle standards, practices, and guidelines for development of CI (configuration item) and non-CI system software.

Documentation: A collection and management of documents identifying plans, processes, and procedures.

Drawing: A computer depiction of graphics or a manually prepared graphic representation of a part or product.

File Name: A term given by the software designer to a specific collection of data.

Firmware: Computer programs and data loaded in a class memory of hardware that contains the software.

First-Article Inspection: The inspection performed to ensure software engineering requirements and processes have been applied to acceptance testing and delivery to customers.

Formal Testing: The process of conducting testing activities and reporting the results in accordance with approved test plans.

Functional Configuration Audit: Prerequisite to acceptance of the configuration item. A technical understanding is accomplished concerning the validation and verification per the test plan concerning software.

Hardware: Physical equipment used in data processing, as compared to computer programs, plans, procedures, and associated documentation.

Implementation Phase: The period of time in the software life cycle during which software work products are created from design documentation.

Informal Engineering Build: Build that is performed with no formal authorization and coordination is applied with teams to ensure the informal engineering build environment is set up.

Inspection: A formal evaluation in which software requirements, designs, or codes are examined in detail to detect faults, violations of development standards, and other problems.

Institute of Electrical and Electronics Engineers (IEEE): Accredited by ANSI (American National Standards Institute) standards.

Integration Testing: An orderly progression of testing which elements of software and hardware are combined and tested.

Interface Requirement: A requirement that specifies a hardware, software, and database in which a system must interface.

Item: An element of a set of data, such as digits, bits, or characteristics, that is treated as a unit.

Modification: A change to software and the process for that change.

Nondevelopment Item: Software used to assist in the development of the deliverable work products, but is not identified as a deliverable product.

Object Code: The output from a compiler directly executable by the computer system.

Peer Review: An important part of verification and a proven mechanism for effective defect removal.

Physical Configuration Audit (PCA): Identifies the product baseline for production and acceptance of the work product audited. PCA verifies that the "as-built" configuration correlates with the "as-designed" product configuration, and the acceptance test requirements are comprehensive and meet the necessary requirements for acceptance of the production unit.

Preliminary Design Review: A formal meeting at which a preliminary design is presented to the user or customer for comment and approval.

Procedure: The documented description of a course of action taken to perform activities or resolve problems; manual steps or processes to be followed.

Process: To perform to defined instructions during the software design/development life cycle.

Product Team (PT): The product team is accountable to management and is composed of members from the appropriate functional disciplines (e.g., engineering, subcontractor, management, product support, etc.) necessary to accomplish day-to-day activities.

Program: A schedule or plan that specifies actions to be taken.

Project Plan: A management approach that describes the work to be done, resources required, methods to be used, reviews, audits, the configuration management, quality assurance procedures to be implemented.

Qualification Testing: Formal testing conducted by the developer for the customer to demonstrate that the software meets specified requirements.

Quality: The totality of features and characteristics of a product or service that has the ability to satisfy required needs.

Quality Assurance: A planned and systematic approach to provide adequate confidence that the product conforms to established requirements.

Quality Management System: Software industries and software programs that establish, document, implement, and maintain effective quality management and continually improve its effectiveness.

Quality Metrics: Measurement of the degrees to which software possesses given attributes that affect quality.

Rational ClearCase: Software tool; an object-oriented database utility used to establish software product archiving, automation, identification, version/change control, software building, product releases, status accounting, and auditing activities.

Rational ClearQuest: Database utility; used for recording, tracking and reporting software work product reports and changes and providing internal access control mechanisms.

Requirement: A condition or capability needed by a user to solve a problem or achieve an objective. The condition of capability must be met by a system to satisfy a contract, standard, or specification.

Requirement Analysis: The process of studying user needs and arriving at a definition of system or software requirements. Verification is also performed for systems and software requirements.

Requirements Phase: The period of time in the software life cycle during which the requirements of a software product, such as functional and performance capabilities, are defined.

Review: Informal or formal review of system requirements, software design, software configuration management, software quality, test, and required data to show compliance to documented plans, processes, and procedures.

Review Board: Established for the software product teams to review and make a disposition of changes that affect controlled software and related documentation.

Risk Management: Process to identify risks and identify an approach to prevent future risks.

Software: Computer programs, procedures, rules, and any documentation pertaining to the operation of data-processing systems. It is in contrast to hardware.

Software Configuration Management: Establishes and maintains the work product identification process and controls changes to identified software work products and their related documentation.

Records and reports information needed to manage software work products effectively, including the status of proposed changes and the implementation status of approved changes. Maintains auditable records of all applicable software work products that help verify conformance to specifications, interface control documents, contract requirements, and as-built software configurations.

Software Contract: Processes and procedures supporting the software work product defined by a purchase contract and technical areas of software design development.

Software Design/Development Process: The process by which a user's needs are translated into software requirements and transformed into design/code being tested, documented, and certified for operational use.

Software Development Facilities: Facilities used for the preparation of software work products prior to delivery to the software and system integration environment or a higher level for testing capabilities.

Software Documentation: Technical data or information that describing or specifying the design or details, explaining the capabilities, and providing instructions for using software.

Software Engineering: A systematic approach to the development, operation, and maintenance of software design/development.

Software Engineering Institute: Resources for improving management practices for addressing software and disciplines that affect software.

Software Life Cycle: The period of time that begins with the decision to develop a software product and ends when the product is delivered.

Software Maintenance: Modification of a software product after delivery.

Software Product: A software entity designated for delivery to the user.

Software Quality: Features and characteristics of a software product that satisfy needs and conform to specifications.

Software/Systems Integration Environment: The primary facility for hardware, software integration, and system-level testing; could include the production of firmware.

Software/Systems Integration Plan (SSIP): Defines or references processes and procedures used to integrate defined work products, systems, or subsystems into a software and systems integration environment.

Software Tools: Computer tools used to develop, test, analyze, and maintain a computer program and its documentation.

Source Code: Computer programs written in a computer language that requires translation by a computer system.

Subcontractor: A company or military and aerospace contractor under a written contract with customers to produce software, hardware, and firmware work products required by contractual requirements.

Subcontractor Plan: Plan for subcontractors to provide required and necessary support to customers per specified requirements in production of work products.

Subcontractor Requirements List: Tracks specification control documents, subcontractor's design, approvals, and acceptance.

Subsystem: A group of assemblies, components, or both combined to perform a single function.

Systems Engineering: Analysis, requirement understanding, and the importance of software design capabilities. Interfaces are defined externally and internally to ensure hardware and software are compatible in supporting team activities.

Testing: The process of exercising or evaluating a system by manual or automated means to verify that requirements satisfy expected results.

Test Readiness Meetings: Ensure that the software tests are complete and carry out the intent of the software test plan and that software to be tested is under formal control and ready for test.

Test Report: A document describing the conduct and results of testing carried out for a system or system component.

Unified Change Management: The approach to manage change in software and systems developments starting from systems design to delivery.

Validation: Demonstrates that the product, as provided, fulfills its intended use.

Verification: Addresses whether the work product properly reflects the specified requirements.

Version Control Document: Identifies and describes a software version consisting of one or more computer software work products; used to release, track, and control software versions.

Version Object Base: A repository for storing software versions of file elements, directories, and data.

Waiver: A written authorization to accept a software configuration item or other designated item that, during production or having been submitted for inspection, is found to depart from specified requirements but is nevertheless considered suitable for use as is or after rework by an approved software method.

Work Product: A product provided by software design that consists of requirements, code, diagrams, documentation, and development folders.

Appendix B: Software/ Systems Integration Plan

Software/Systems Integration Plan (SSIP)

PLAN NUMBER: RELEASE/REVISION: RELEASE/REVISION DATE:

Assigned Plan Number   NEW   mm/dd/yyyy

OWNER:

Program or Project

Plan Information

Plan Type	Revised New Release Date	Contract Number
Formal/revision	dd/mm/yyyy	TBD or N/A

Signatures:

Author: _____*Signature*_____   _____*Program/Project*_____   _____*mm/dd/yyyy*_____
                        Name

Check by: _____*Signature*_____   _____*Program/Project*_____   _____*mm/dd/yyyy*_____
                        Name

Approved by: _____*Signature*_____   _____*Program/Project*_____   _____*mm/dd/yyyy*_____
                        Name

Released by: _____*Signature*_____   _____*Program/Project*_____   _____*mm/dd/yyyy*_____
                        Name

ABSTRACT

Example:

The *program or project* software/systems integration plan (SSIP) is the document for defining plans, processes, and procedures for software/systems work product-level test and evaluation. The *program or project* consists of computing hardware and software. The software consists of an operating system and application. The hardware consists of computers, displays, network interfaces, and interfaces to other subsystems. This SSIP describes the test environment to be used for the testing, identifies the tests to be performed, and provides an overview for test activities.

KEYWORDS

Development File Folder (DFF)
Development Lab (DL)
Development Plan (DP)
Quality Assurance (QA)
Systems Integration Facility (SIF)
Software Configuration Management (SCM)
Software Engineering Institute (SEI)
Software Systems Integration Plan (SSIP)

CONTENTS

1 SSIP PLAN OVERVIEW

This software systems integration plan (SSIP) is a document for defining plans, processes, and procedures for the integration of software and systems for high-level developmental testing for *programs* or *projects.*

The processes used are the following: integration definition and development, integration procedure development reviews, integration activity execution, and integration evaluation and reporting. The SSIP also includes overall integration planning and coordination with other test activities, risk assessment, product evaluation, software configuration management, and other related software support activities.

This plan is a working document that will be revised as necessary throughout the *program or project* software development life cycle. The software development life-cycle plan, schedules, and the development plan (DP) will be evaluated if required at the beginning of each of the software builds. This working plan is not a deliverable to the customer but will be provided along with the development file folder (DFF) required.

The SSIP is prepared using guidance from MIL-STD-xxx data item description (DID) and the Institute of Electrical and Electronics Engineers (IEEE) standard. The methods, standards, and procedures described here were derived from the level 3 processes of the Software Engineering Institute (SEI) Capability Maturity Model Integration (CMMI®). The SSIP identifies the required organizations, resources, and schedules for integration activities.

2 SENIOR MANAGEMENT

The *program* or *project* senior management function is to perform guidance, support, monitor, and report progress regarding software and systems integration to teams and define the role and task for software- and systems-related activities as shown in Table B.1.

2.1 Test Schedules

Software development will follow the *program* or *project* master schedules. The master schedule defines the dates that serve as the milestones for software and systems integration activities.

TABLE B.1

Roles and Responsibilities

Group	Roles	Task
Test team	Trained testers	Software/system test activities
Software team	Software design	Software-related activities
Work product teams	Reviewers	Participate in reviews of software design and test

2.2 Software Tools

The software and systems integration environment encompasses equipment, location, and tools used to develop and test facility environments. Software tools are used to maintain the integrity and structure of the existing software development and applicable documentation.

Note: Such tools are, for instance, ClearCase and ClearQuest.

2.3 Relationship to Other Plans

The relationship to other plans increases the objectivity for the *program* or *project* to ensure better oversight with the capabilities to reference plans that support the software and systems integration activities.

Note: List plans.

Development plans
Configuration management plans
Quality plans

3 TEST

The conduct and completion of the test provide verification to ensure that *programs* or *projects* meet the requirements of the software and system design.

3.1 Test Products

Inputs for test products to support software and system integration include:

- Software requirements
- Approved changes
- Acquired engineering products
- Software resources

- Controlled software
- Authorization for updates to test products

Plans, procedures, and data include:

- The SSIP
- Software/system integration test procedures
- High-level test plans
- Problem reports
- Software/system integration measurement data

3.1.1 Software

Defined and documented software plans and procedures include:

- Development plan (DP)
- Software user manuals (SUMs)
- Software configuration management plan (SCMP)
- Software build and installation procedures
- Software quality plan (SQP)
- Integration facilities operational plans and procedures

3.1.2 Systems

Defined and documented system design plans and procedures include:

- Systems engineering plan (SEP)
- Systems design procedures
- Systems test reports
- Hardware drawings
- Integration facility configurations
- Hardware serial number verification procedures or instructions

3.1.3 Hardware

Defined hardware includes:

- Workstations
- Display units
- Printers

- Disk sets
- Drivers and interface cards
- Networks or servers

3.1.4 Test Documentation

Defined test documentation includes:

- Test plan (TP)
- High-level test plans
- Test report (TR)
- Installation procedures

4 TEST APPROACH

The test approach for *programs* or *projects* is an integrated plan and activity to begin preparation of informal and formal test plans and procedures based on the specifications listed in Table B.2.

4.1 Informal Test

To establish controls and reporting processes for informal use, the *program* or *project* approach is based on informal and draft documents. Results of informal engineering testing of the internal documents and

TABLE B.2

Test Approach

Test Team Focus	Test Approach
Prepares for incremental software and systems integration of functional test methods	Incremental test methods provide visibility into the evolving work product and reduce common problems of delayed visibility late in software and system integration and testing.
Mature test processes and tools	Software test tools are available tools that automate and integrate the software development and integration.
Upgrades the path to meet future issues and utilize advanced technologies	Maximum utilization of commercial software and hardware.

artifacts (e.g., version control documents [VCDs], test plans, test procedures, data, metrics, etc.) are not released to customers and are maintained internally for checkout, troubleshooting, and recommendations to
start the formal test.

4.2 Formal Test

The *program* or *project* formal test is conducted to official and formal
released procedures, plans, and instructions. The formal test states per
high-level authorization and documentation the objective and success criteria. The formal test requirement describes the configuration of
the software or system item under test and lists the test equipment and
required support. Formal engineering testing of the internal documents
and artifacts (e.g., VCDs, test plans, test procedures, data, metrics, etc.) are
released to customers and maintained internally.

5 RESPONSIBILITIES

The responsibilities of systems and software engineering, software configuration management (SCM), software quality organization, and the test
team require *program* or *project* support during the software and systems
integration activities stated in Figure B.1.

5.1 Systems Engineering

Technical requirements are provided by systems engineering personnel
to allocate technical requirements for programs or projects that consist of
both software and hardware.

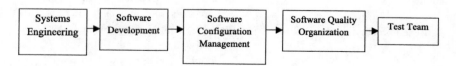

FIGURE B.1
Software team responsibilities.

5.2 Software Development

Software design personnel provide software development support during software and systems integration testing in a development lab (DL) and integration facilities. They are responsible for troubleshooting, resolving software problems, and supporting isolation of system problems that occur while testing hardware and software.

5.3 Software Configuration Management

The software configuration management (SCM) team is responsible for building and providing configuration control of software for test organizations to use in the DL and integration facility. For software and integration testing, the SCM team is responsible for the configuration of the software in conjunction with other software build teams. The software builds and installation in the required facilities are performed by SCM following version control document (VCD) installation instructions or installation procedures.

5.4 Software Quality Organization

The software quality (SQ) organization will support system and software tests conducted as defined in the software quality plan (SQP). The hardware quality assurance (QA) personnel are not required to support the test but monitor execution for software during integration activities.

5.5 Test Team

The *program* or *project Test Team* is responsible for the conduct and completion of informal and formal test activities per Figure B.2 to ensure a level of functional maturity. The test team has the overall responsibility to plan, schedule, and conduct test activities in the integration facilities.

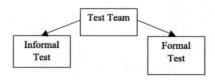

FIGURE B.2
Informal and formal tests.

6 FACILITIES OPERATION

The *program* or *project* facility operation environment provides integration for hardware and software and systems integration to support software design and equipment integration. The facility operation will be used to integrate and test the work products and build up and support incremental deliveries of software builds.

6.1 Metrics

Metrics are used on the *program* or *project* to manage facility operations activities. These metrics are used to evaluate the maturity of the software, measure progress of development, test efforts, and identify software risks during integration in an integration environment.

6.2 Risk Management

Risk management describes the process used by *program* or *projects* to identify and monitor facility operation risks inside software and system integration activities.

7 NOTES

Industry standards such as MIL-STD-xxx are used as a baseline for *programs* and *projects* and developing methodology for integration and testing. The IEEE-xxx standard was used as guidance for identifying integration and test methods.

8 ACRONYMS

DP: development plan
SSIP: software/systems integration plan
VCD: version control document
(*Continue listing acronyms, i.e., DFF, development file folder*).

9 FIGURES

Figure B.1. Software life-cycle responsibilities.
Figure B.2. Informal and formal test.

10 TABLES

Table B.1. Roles and Responsibilities
Table B.2. Test Approach

Appendix C: Software Audit Checklist

Software Audit Checklist

Subcontractor Company Name

VCD Part Number and Rev Level: ___
Deliverable: *Definition and Functions*
Participants: *List Names*

Customer Representative: *Name*
Start Date: *mm/dd/yyyy*
Completion Date: *mm/dd/yyyy*

Subcontractor Manager: *Name*　　　Customer Manager: *Name*

Sub = Subcontractor, Cust = Customer, Yes = Y, No = N

Verifications	Sub Y/N	Cust Y/N	Condition Noted
Readiness review: • Plans/procedures released • Software environments available • Personnel prepared and available • Software configured for test			*Start Date: mm/dd/yyyy*
Development and testing conducted			*Ensure this development and testing phase is completed to support the software audit.*
Objectives: • Functional requirements satisfied • Problems with hardware or software identified and recorded			*N = Action Item (AI#1)* *Y = Pass*
Software configuration management: • VCD released • Software configuration management plan *revision* • Plans and procedures released • Software media controlled • Required software installed for verification and validation			*Notes:*

Continued

Verifications	Sub Y/N	Cust Y/N	Condition Noted
Configuration status accounting report: • Operation procedures released • High-level changes approved • Plans in database management • Procedures in database management • Documents and drawings released • Quality buy-offs	N		*AI#1*
Test conduct: • Test procedures performed in accordance with company policies • Paperwork authorized for test • Test failures recorded • Configuration issues recorded • As-run procedures performed • Test reports released/*revision* • Test complete			*Notes:*
Data package: Software deliverables prepared for retention in a controlled computer media library (CML) and archived both on-site and off-site			*Notes:*

Test Conduct	Sub Y/N	Cust Y/N	Condition Noted
Test procedures:			*Notes:*
• Test data procedure *revision*			
• Operating procedures *revision*			
• Software user manual *revision*			
• Software development plan *revision*			
• Software user manual *revision*			
• Test report *revision*			

Product Release and Acceptance	Sub Y/N	Cust Y/N	Condition Noted
Any regression testing of any audited test steps recorded.	N		*AI#2*
Test procedures released and under change control.			*Notes:*

Product Release and Acceptance	Sub Y/N	Cust Y/N	Condition Noted
Test reports released and under change control.	N		*AI#3*
Top-level assembly or outline drawings have been reviewed and approved by customer.		N	*AI#1*
Test environment was defined and controlled.			*Notes:*
Open and closed: • Problem reports • Corrective actions • Issues • Findings			*Notes:*

Software Audit Completion

Software Audit Completion	Sub Y/N	Cust Y/N	Condition Noted
Is there evidence of software acceptance?			*Notes:*
Are completion dates for any open action items defined?			*Notes:*

Software Audit Action Items

Description	ECD	Complete
AI#1: Document	mm/dd/yyyy	
AI#2: Document	mm/dd/yyyy	
AI#3: Document	mm/dd/yyyy	

Customers Action Items

Description	ECD	Complete
AI#1: Document	mm/dd/yyyy	

Team Participation:

- Program or project managers
- Systems engineering
- Software design
- Software test
- Configuratiron management
- Quality assurance

Software Audit was conducted and performed at: *Facility Name*

Date Completed: *mm/dd/yyyy*

The current Software Audit status:

☐ Closed as acceptable with no action items.
☐ Closed as acceptable with action items.
☐ Open pending completion of defined action items.

Name	*mm/dd/yyyy*
Customer Representative	Date

Appendix D: Software Checklist PCA

< EXAMPLE >

<u>Software PCA Checklist</u>

Program or Project
Integration Facility Location

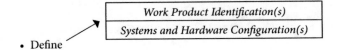

- Define

| Work Product Identification(s) |
| Systems and Hardware Configuration(s) |

Plans:

Document (1)	Document Number/Rev		Yes	No
Quality plan (QP)	Nnnn-nnnnn-n	A	X	
Systems engineering plan (SEP)	Nnnn-nnnnn-n	B	X	
Software configuration management plan (SCMP)	Nnnn-nnnnn-n	C		X
Software quality plan (SQP)	Nnnn-nnnnn-n	D	X	
Development plan (DP)	Nnnn-nnnnn-n	E	X	
Configuration management plan (CMP)	Nnnn-nnnnn-n	F		
Configuration audit plan (CAP)	Nnnn-nnnnn-n	G		X
Software subcontractor plan (SSP)	Nnnn-nnnnn-n	H		X
Software systems integration plan (SSIP)	Nnnn-nnnnn-n	J		
Test plan (TP)	Nnnn-nnnnn-n	K		X
Computer media library (CML) plan	Nnnn-nnnnn-n	L	X	

1. Additional Plans: Design Engineer/Test Engineer Generated

Procedures:

Procedure (2)	Document Number/Rev		Yes	No
Software instructions	Nnnn-nnnnn-n	A	X	
Facility setup procedures	Nnnn-nnnnn-n	B	X	
Build procedures	Nnnn-nnnnn-n	C		X
Version control document (VCD)	Nnnn-nnnnn-n	D	X	
Integration test procedures	Nnnn-nnnnn-n	E	X	
Integration test reports	Nnnn-nnnnn-n	F		X
Software loading instructions	Nnnn-nnnnn-n	H	X	
Software logs (informal integration)	Nnnn-nnnnn-n	J		X
Work product authorizing procedures	Nnnn-nnnnn-n	K		X

2. Additional Corrective Actions: Quality Generated

Records:

Problem Reports (1)	Description	Open	Closed
N-nnnn-01	Test failures	X	
N-nnnn-02	Integration errors	X	
N-nnnn-03	Troubleshoot integration problems		X
N-nnnn-04	Software design problems		X
N-nnnn-05	Steps missing in procedures	X	
N-nnnn-06	Coding errors		X
N-nnnn-07	System installations incorrect	X	
N-nnnn-08	Software build errors	X	
Corrective Actions (2)	**Description**	**Open**	**Closed**
CAnnnn-01	Process not compliant during code development	X	
CAnnnn-02	Monitor formal test activities		X
CAnnnn-03	Project processes not followed		X
CAnnnn-04	Implementation process concerns	X	
CAnnnn-05	Plans not followed during test		X
CAnnnn-06	Plans and procedures not released	X	

3. Additional Problem Reports: Design Engineer/Test Engineer Generated

4. Additional Corrective Actions: Quality Generated

Computer Media Library	Comments
Commercial off the shelf (COTS)	Part number, version, title
Nondevelopment item (NDI	Part number, version, title
Company developed software	Part number, version, title
Computer media library (CML) plan	Part number, version, title
Vendor installation procedures	Documentation

Computer Media Library (Procedure): Incoming and Outgoing Media (Audit)

Attachment: Computer Media Library (CML) Log sheet
See below

ID	Source	P/N	Ver	Qty	Title	Media	Date	Vendor	Location
1	COTS	nnnnn	v001	2	PC apps	CD	mm/dd/yy	Name	CML1
2	COTS	nnnnn	v003	1	Unix apps	DVD	mm/dd/yy	Name	CML2
3	NDI	nnnnn	v005	4	MS Word	CD	mm/dd/yy	Name	CML3
4	Company	nnnnn	v007	2	Product	CD	mm/dd/yy	Name	CML4
5	COTS	nnnnn	v009	2	PC apps	DVD	mm/dd/yy	Name	CML5
6	NDI	N/A	N/A	1	Instruction	Doc	mm/dd/yy	Name	CML6

PCA Checklist:

Activities	Comments
• PCA entry criteria accomplished: ✓ Readiness review successfully held ✓ Defined responsibilities and authority ✓ Agreed-to agenda ✓ Presentation including scope and in-brief materials ✓ Presentation material clear and sufficient in detail and consistent within scope	Date Started: mm/dd/yyyy
• Product configuration baseline identified. Were the operating and software support documents reviewed (VCD, TPs, TRs, etc.)? Installation software identified in VCDs with reference to media and systems	

Continued

Activities	Comments
• Specification review and validation to define the configuration item, testing, mobility/transportability, and packaging requirements: ✓ Packaging plan/requirements review complete ✓ Test procedures and results complete	
• Documents review against as-built and variations, including outstanding design changes, part numbers, and description: ✓ Changes incorporated between test baseline and release baseline? ✓ Installation and inspections complete and closed	
• Review of unincorporated design changes. ✓ Outstanding changes, problem reports	
• Review waivers and deviations to specifications and standards: ✓ Action item database open items impacting software	
• Document release system for control of processing and formal release of engineering changes: ✓ Test tools identified? ✓ Software tools identified ✓ Software part number ✓ Review of build processes ✓ Software media labeling requirements ✓ Software media storage requirement	
• Was the software end item reviewed? 　✓ Software design descriptions 　✓ Software requirements ✓ Status of reviews (informal/formal) ✓ Findings/status of quality evaluations/reviews/audits ✓ Subcontractors developed software ✓ Embedded COTS/NDI/company developed	
• Exit criteria: ✓ Roster ✓ Agenda and presentation data ✓ Action item log and action items ✓ Data package ✓ Signed certification sheets ✓ Test data is complete and accurate	Date Completed: mm/dd/yyyy

Certifications:

Data/Media	Status
PCA cover sheet	Open
Certification package scope/purpose	Closed
Product configuration baseline established	Closed
Specification reviews and validation	Open
Software drawing/document review	Closed
Unincorporated software changes	Closed
Software deviations/waivers	Open
Computer media library (CML)	Closed

- *Status* → *OPEN*
→ *CLOSED*

Action Items:

AI Number	Description	ECD
AI-001	Problem reports open	mm/dd/yyyy
AI-002	Corrective actions open	mm/dd/yyyy
AI-003	Build procedures not released	mm/dd/yyyy
AI-004	SCMP not available	mm/dd/yyyy
AI-005	Locate software test logs	mm/dd/yyyy
AI-006	Testing not complete before PCA	mm/dd/yyyy

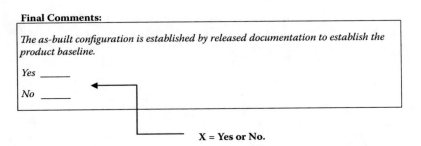

Final Comments:

The as-built configuration is established by released documentation to establish the product baseline.

Yes _____

No _____

X = Yes or No.

Company Representative(s)   mm/dd/yyyy

Index

Printed by Publishers' Graphics Kentucky